Ecology
and Applied
Environmental
Science

Ecology and Applied Environmental Science

Kimon Hadjibiros

CRC Press
Taylor & Francis Group
Boca Raton   London   New York

CRC Press is an imprint of the
Taylor & Francis Group, an **informa** business

CRC Press
Taylor & Francis Group
6000 Broken Sound Parkway NW, Suite 300
Boca Raton, FL 33487-2742

First issued in paperback 2019

© 2014 by Taylor & Francis Group, LLC
CRC Press is an imprint of Taylor & Francis Group, an Informa business

No claim to original U.S. Government works

ISBN-13: 978-1-4665-7009-2 (hbk)
ISBN-13: 978-0-367-86776-8 (pbk)

DOI: 10.1201/b14609

Visit the Taylor & Francis Web site at
http://www.taylorandfrancis.com

and the CRC Press Web site at
http://www.crcpress.com

To Mata and the earthly manifestations of universal forms

Contents

Preface

Systems of various sizes, forms, and synthesis exist at all spatial and temporal scales of the universe. When they tend to increase, an opposite tendency that prevents this increase beyond a certain limit may be manifested; a perpetual tug-of-war between the two tendencies, the impulse and the constraint, determines the system's situation. This kind of opposition also emerges between development and the environment in the modern world. The endless quest for more or better, based on the view that man's continuous expansion is possible, is manifested by the pursuit of economic, social, or cultural development and corresponds to the impulse. The opposite tendency, constraint, corresponds to the will for conservation, frugality, or the prevalent concern about environmental threats and technological hazards. This book presents a composition of scientific and technological, environmental, and ecological issues that could contribute to a timely political and social reflection. It is dedicated to George Rossis (1948–1994), physicist and philosopher, who furthered the analysis of fundamental concepts through fruitful discussions.

I would like to offer my special thanks to all the people who have contributed to the preparation of this book, especially to Rena Karakatsani and Tasia Sialiaridou who took care of the English text and the figures, respectively; to Sofia Makrogianni who undertook the correspondence related to rights and permissions; as well as to the production staff of Taylor and Francis, and especially to Tony Moore, acquiring editor at Taylor and Francis, for his valuable help and encouragement.

Kimon Hadjibiros

Introduction

ABOUT ECOLOGY

Ecology is the science that studies the relationships between organisms and the environment. It was created as an extension of biology but it is also based on chemistry and physics, and it uses mathematics and information sciences to a significant degree. The concept of ecology has expanded in the last several decades, acquiring social and political dimensions due to the serious environmental problems of the modern world, their direct or indirect relation with the process of economic development, and the increasing relative social sensitization.

The main volume of research activities in the field of scientific ecology is still developing around the axis of environmental biology, seeking to answer basic questions about the distribution and abundance of organisms on the planet. These questions attract a keen scientific interest from the research community, but they are only partly related to issues of perturbation and degradation, damage caused by human activities, and environmental protection or restoration. However, the science of ecology can help us approach these issues not only through the study of the organisms' role but even more through the examination of environmental processes from other viewpoints, such as the role of energy flow and circulation of chemical elements in the ecosystems' structure, function, and stability. Systems ecology bypasses the details and focuses on basic phenomena, with the exploration of nature's general function as its primary goal. Its expectation is that the analysis of basic processes will offer the possibility for more sound environmental management and for coping with the related problems. At the same time, the achievement of a better understanding of basic processes contributes to raise the awareness of the agencies that intervene in the environment, promoting the development of the necessary ecological conscience.

The object of systems ecology is the study of ecosystems. The concept of ecosystem is primary in this consideration and promotes theoretical and applied research as well as global understanding of environmental problems by the lay people. Barry Commoner's four laws of popularized ecology—"Nature knows better," "Everything is connected with everything else," "Everything must go somewhere," and "There is no such thing as a free lunch"—are characteristic, and essentially they refer to the concept of ecosystem.

The ecosystem can be defined as a finite total of organisms that occupies a particular place and has an appointed structure; between this and the physicochemical environment there are mutual influences and exchanges of matter and energy that contribute to the creation of an organized functional unity. Microorganisms, plants, and animals constitute the biological part of the ecosystem. The concept of physicochemical environment covers every kind of matter, inorganic or organic, that is not alive, as well as every physical factor, like radiation or the wind. A lake, a forest, an ocean, and even an aquarium can be examined as an ecosystem. The total of ecosystems, which occupies a relatively thin layer above and under the surface of the Earth, constitutes the planetary ecosystem that is called biosphere.

Man perturbs and alters the ecosystems and has done so since the dawn of civilization. In the last few decades, however, economic growth and the strengthening of technology have expanded the magnitude of environmental changes, their extent in space, and their persistence in time. The favourable results of the changes on man's prosperity are many, but interventions on ecosystems can have significant negative consequences for the quality of the environment, the biosphere's equilibrium, and for human well-being. The technological abilities for dealing with environmental problems are significant, but they have their limits. Consequently, the imposition of rules and constraints on economic development is necessary, and they should result from the scientific study of the environment and of the ecosystems in particular. Understanding these issues is especially important for the operator of technology and especially the engineer who is the main agent of human intervention in the environment. Infrastructure works, such as dams, irrigation or drainage networks, roads, ports and airports, industrial and other productive activities such as power stations or mines, the various solid, liquid, or gaseous waste, etc., are classic causes of environmental degradation. It is useful for every scientist, engineer, manager, or decision maker to have a minimum of knowledge about the functions of ecosystems and the perturbations caused to the environment by man's intervention. Integrated knowledge is conducive in the achievement of productive and developmental goals with the least possible intervention and the least environmental cost. It also furthers awareness and sensitization so that the useless wounding of the landscape can be avoided and the natural and cultural elements of every place may be respected.

ABOUT THE BOOK

This book attempts to link contemporary environmental problems with the main principles of scientific ecology. The understanding of problems and the explanation of phenomena are better served if the lay reader becomes familiarized with the organization of ecological systems at an ecosystem level rather than at a population level. Thus, in contrast to the international trends of most scientific books that bear the term "Ecology" in their titles, the approach that was chosen here places a greater emphasis on the physico-chemical phenomena of ecosystems than on the biological functions that define the organisms' relationship with their environment.

The effort for a presentation that would also be accessible to the wider public has limited and simplified—but cannot abolish—the mathematical approaches. In the case of systems ecology, mathematics is necessary for the codification of the subjects and the setting off of the critical parameters that determine the ecosystems' behaviour. Special emphasis has been placed on examination of fundamental concepts, such as energy flow, competition, predation, or limiting factor. Familiarization with these concepts contributes to an integrated consideration, and therefore to a deeper understanding and interpretation of ecological phenomena. Certain subjects, though, have secondary significance or constitute more detailed approaches and are more difficult to understand. Their elimination would deprive the book of necessary documentation. However, their omission by the reader does not have negative consequences for understanding, while it gives the reader the possibility to focus on the more significant subjects and run through the book's material more quickly without losing a great part of the information.

The book consists of three parts. Part 1 presents basic knowledge of ecological science on the basis of the systemic approach. Chapter 1 introduces basic concepts of systems theory, the model method, and the concept of ecosystem. Chapter 2 examines the organisms' categories in relation to their role in the ecosystems, as well as the basic physical and chemical processes that affect them. Chapter 3 refers to the structure and the changes of populations and to analysis of their major mutual influences, using the necessary mathematical models; also, information is given about the changes of human populations. Chapter 4 elaborates on the roles of primary production, limiting factors, and the circulation of chemical substances and energy flow in the ecosystems, and it presents the basic functions of aquatic ecosystems.

The most important issues related with environmental degradation by human activities are the focus of Part 2 and includes a brief survey of the planet's ecosystems and the alterations of the natural environment, e.g. forests, coasts, wetlands, soil, biodiversity, and the natural landscape; special urban environmental problems; environmental degradation due to toxic pollution, including radioactivity; various water pollution phenomena such as deoxygenation and eutrophication; and various forms of atmospheric alterations such as photochemical pollution, ozone layer depletion, and the greenhouse effect.

The environmental management prospects are the focus of the Part 3. The four chapters discuss the potential for technological control of pollution from liquid, solid, and gaseous waste; existing tools for environmental management; ecological and social aspects of biodiversity and landscape protection; and the antithesis between development and environment in combination with ideas about sustainability.

Attention has been given to the clarification of issues about which dubious views are often held by the public, including:

- The confusion between harmful effects or toxicity from radioactivity, from ultraviolet radiation, from inorganic or organic chemical pollutants
- The easy attribution of many environmental damages to toxic pollution
- The underestimation of ecological perturbations not related to chemical pollution
- The decisive role of the limiting factors in ecosystems
- The biomass and oxygen produced by the land and marine ecosystems
- The confusion about the positive and negative effects of ozone
- The role of algae in the aquatic ecosystems and the naturalness or not naturalness of eutrophication
- The supposed oxygen deficiency in the cities and more generally in the atmosphere, the oxygen production by the forests or urban green and the risk of oxygen depletion in the water environment

Part I

Ecological Systems

Chapter 1

Systems and Ecosystems

DOI: 10.1201/b14609-1

1.1 THE CONCEPT OF SYSTEM

The term *system* is used in many different contexts. Reference is made, inter alia, to natural, organic, social, ideological, political, cultural, religious, administrative, legislative, monetary, urban-planning, industrial, agricultural, insurance, healthcare, education, power, and computer and communications systems.

The scientific use of the term comes from physics, and more specifically from thermodynamics, where a system is defined as the portion of the universe that we are considering, in contrast to the environment, which is the rest of the universe. The widespread use of systems in many sciences, as far back as 1950, was based on General Systems Theory, a separate branch of science that includes a set of mathematical methods. Its purpose is to search for the general properties and features of the behaviour of systems. It was founded by Canadian theoretical biologist Ludwig von Bertalanffy. Analogous to General Systems Theory is the approach of cybernetics, which is defined as the science of control and communication in animals and machines. It was developed by American mathematician Norbert Wiener in 1948.

One definition of a system is a finite set of interacting parts, which can be considered to be a unit. It should be noted that:

- A system has limits; it cannot be infinite.
- It consists of different parts, which may be concrete or abstract, simple or complex.
- There must be interaction among its parts. Some authors require that such interaction be non-repetitive or non-linear.
- One prerequisite is for the system to create a new entity, at a level of description different from the level where its parts are described. This means that the connection among its parts is not merely cumulative but creative, i.e. it produces new features and properties (Figure 1.1).

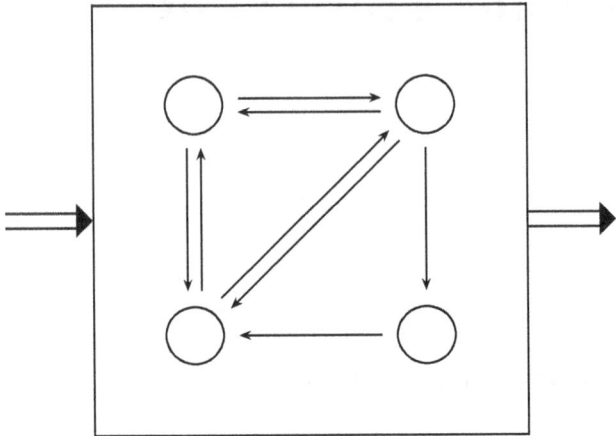

Figure 1.1 Four-part system. (From Hadjibiros, K. (2007). *Ecology. Ecosystems and Environmental Protection,* 3rd edition. Symmetria, Athens (in Greek). With permission.)

The concept of system contributes to a theoretical approach to the organization of the differences observed in various description levels of the world, such as physicochemical, biological, psychological, and social. The laws of each lower level, e.g. of physics and chemistry, apply and are totally compatible with the phenomena of the higher levels of description but do not contain all the concepts necessary for an understanding of these phenomena. The extensive use of systems in biology, in ecology, and in the social sciences is due to the complexity of the relevant phenomena. This consideration stresses the organisation of the subject rather than its material structure. Some of the characteristic issues it examines are organisation, differentiation, rank, information, communication, and control. The study of problems on the basis of the relationship between the interaction of the parts and the whole is defined as a systemic approach.

1.2 CONCEPT AND USE OF MODELS

The model can be defined as a simplified representation of reality. This broad definition includes many material and intellectual objects: for example, every word or every mental image is a model. In the exact sciences the model is used as a basic tool for reconstructing nature in order to study it. In this sense it is a hypothetical system that is realistic enough to meet the objective of study.

The use of models in science has two different aims: to explain and to forecast phenomena. Two types of models correspond to these aims:

- Tactical models, which are more empirical and are applied mainly in specific instances, aiming at forecasting. They usually have a practical purpose, are based on a large number of detailed hypotheses, have a relatively narrow scope of application, and lead to quantitative conclusions, mainly of practical value.
- Strategic models, which are more abstract and general, which further the investigation of basic mechanisms and causal relationships governing the functioning of the system, and whose main purpose is to explain. They are usually theoretical in nature, contain fundamental concepts, are based on a few simple hypotheses, are relatively broad in scope, and lead more to qualitative conclusions.

These two categories of models constitute the two extremes of what is essentially a continuous spectrum. Most of the models used in theoretical and practical research belong to some point on the spectrum—that is, they show a mix of features of both categories, depending on the object of study and the opportunities to approach the subject. Models always have the following three properties:

- Generality, or applicability of the model to various situations
- Realism, or a degree of correspondence between the model's logical structure and the reality it is supposed to represent
- Precision, or the model's ability to quantitatively reproduce the reality through its results

A perfect model is a utopia. It is not possible to have a simultaneous maximisation of generality, realism, and precision. Maximisation of these three properties in the right combination determines the optimum use of models, depending on the type and objective of the approach. To use the example of ecological models, as a rule those usually used by biologists or other laboratory scientists sacrifice generality for the sake of precision and realism, whereas those proposed by scientists with a background in mathematics tend to sacrifice realism for the sake of generality and precision. What is more, many theoretical ecologists believe that the models creating the theoretical basis of ecology should not overstress precision but lay the emphasis on generality and realism.

1.3 SYSTEM STATE AND STABILITY

System state is a fundamental concept. If we consider that the system is represented (that is, a complete image of it is formed) by n variables in a space of n dimensions, then the total values of n variables at a certain moment in time define the state of the system. The n variables change values over the course of time, forming an equal number of paths in the space of n dimensions. The temporal evolution of the system can be represented by the set of those paths. In order to fully determine this evolution, appropriate variables must be chosen, which will be sufficient to describe the variations of the system with no uncertainty. If the change in the variables is made under certain conditions, then it can be represented by a mathematical equation, usually a differential one. In this case, a set of equations describes the temporal evolution of the state of the system; they are called state equations.

When the state of a system is stagnant, then the system is at an equilibrium point. If the system is at that point at some moment in time and no significant disturbance occurs, it is expected to remain at the same point at the next moment in time. The state of equilibrium is expressed mathematically by the assumption that the temporal derivative of all the system's variables is equal to zero:

$$\frac{dx_i}{dt} = 0, \quad i = 1, 2, \ldots n$$

The temporal evolution of a system can be represented graphically in two ways. In the first case, the change in each variable can be represented by a two-dimensional graph, where the value of the variable is measured on the vertical axis and time on the horizontal axis. Thus the temporal evolution of the system is depicted in a set of such diagrams, which are equal in number with the system variables. The following equations apply:

$$x_i = f_i(t), \quad i = 1, 2, \ldots n$$

In the second case, the change in each variable is represented by a curve in a space of n dimensions, the phase space. The evolution of the system is depicted by a line in a hypersurface of n dimensions where each axis corresponds to one variable, and time t has been eliminated.

Stability is a property of the specific equilibrium point and is defined in relation to it. If the system moves away from the equilibrium point, i.e. undergoes a disturbance and has the tendency to return to the same point, then the specific equilibrium point is stable. By contrast, if a system that has been disturbed continues to move away from the equilibrium point, then that point is unstable.

There are varying degrees of stability, depending on how great a disturbance the equilibrium point can withstand. If the system can return to the equilibrium point only when there have been small deviations from it, then we have local stability. In this case, in the phase space around the equilibrium point A, there will be a specific area, from all points of which the

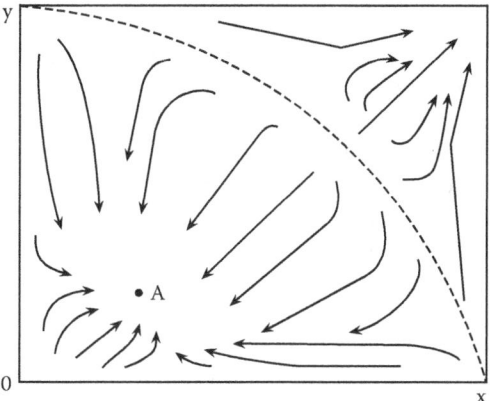

Figure 1.2 The area of attraction of point A is defined by the axes and the broken line. (From Hadjibiros, K. (2007). *Ecology. Ecosystems and Environmental Protection,* 3rd edition. Symmetria, Athens (in Greek). With permission.)

system will have the tendency to move towards A. This area is called the area of attraction of the equilibrium point (Figure 1.2).

If, on the contrary, the return to the equilibrium point occurs in the case of any departure, no matter how great, then we have overall stability. No matter what state the system is in, it will have the tendency to move towards the specific equilibrium point which has the property of overall stability. In such a case, therefore, no other equilibrium point can exist.

The stability of the equilibrium points is analogous to the relief of a landscape with smooth terrain containing hills and depressions. The system is depicted by a rolling sphere and reaches equilibrium when it is at rest. The lowest points of the depressions correspond to stable and the hilltops to unstable equilibrium points. The line of peaks around a depression corresponds to the boundary of its area of attraction.

Interactions of internal and external factors can cause perturbations, leading the system temporarily or permanently away from a state of equilibrium. If a factor, able to overturn the equilibrium, affects it, the system could possibly find a new equilibrium. The system has the property of resilience when it is able to absorb great perturbations.

A system's response to an external perturbation can play an important role in its behavior's regulation. Of special interest is the negative feedback, where an external perturbation causes feedback that may lead the system to equilibrium, serving the control. Negative feedback plays a fundamental role in the regulative processes of many systems. On the contrary, a positive feedback may lead the system to an explosive behavior, that is, out of control.

A more detailed approach to the relevant issues is presented in the Appendix.

1.4 THE ECOSYSTEM AS A SYSTEM

The ecosystem is a complex concept, the understanding and treatment of which are of particular interest but also present many difficulties. It is obvious that the ecosystem has the property of constituting a unit and at the same time a set of elements. Its elements are connected in a variety of ways, via dependences and interactions. Their close correlation forms a unit with its own characteristics, which does not allow the behaviour of the ecosystem to be reduced to a mere sum of the properties of its parts. Thus the ecosystem is an instance where the theoretical concept of the system can be successfully applied to reality.

The action of external factors becomes perceptible by means of one or more of its components, but it affects the ecosystem, to a greater or lesser degree, as a unified whole. By causing actions and chain reactions, it disturbs the ecosystem, causing it to depart for a time from a state of equilibrium, or propelling it to a new equilibrium, or, in extreme cases, driving it to collapse.

Ecosystems are extremely complex systems, with a great number of internal and external interactions. Usually just the trophic relationships among the various organisms of an ecosystem form a complex web of internal links. Human beings are one of the numerous interacting elements of ecosystems. They have acquired the ability to bring about great changes in ecosystems, but not to exercise effective control. Thus they are often unable to manage them successfully and protect them from the undesirable side effects of their activities.

A natural ecosystem is generally maintained in a state of dynamic equilibrium, where various processes of control, regulation, and feedback play a fundamental part. In parallel, if the ecosystem has not yet reached its most mature form, it evolves at a slow pace, through successive states of equilibrium, where the phenomenon of ecological succession may be observed.

The stability of an ecosystem is dependent on the constant, regular flow of energy and circulation of vital chemical substances through its structures. Due to its organisation, when the ecosystem moves away from the state of equilibrium without escaping from certain bounds, internal changes are generated which return it to the state of equilibrium. In certain cases, the ecosystem has this ability (resilience) even for very great disturbances.

The maintenance of states of equilibrium becomes apparent when the long-term persistence of ecosystems is observed, and it constitutes the objective of managing or protecting them from human disturbances. Thus the property that in general serves the sought-for conservation of ecosystems is the stability of equilibrium states.

An examination of the natural environment using the concept of the ecosystem is fruitful in many applications and meets scientific needs, such as:

- Understanding the functioning of parts of the environment and/or the whole biosphere

- Forecasting environmental phenomena or the behaviour of environmental systems
- Studying environmental impacts of works or activities or studying the management of parts of the environment, and studying the environment in general
- Determining the need, the type, and the manner of an environmental intervention

In all the above cases, one of the basic direct or indirect objectives of the approaches is to determine the temporal evolution, the behaviour, the equilibrium points, and the stability or the instability of the systems under study.

1.5 MATHEMATICAL MODELS OF ECOSYSTEMS

Models in ecology are often applied either to population systems or to ecosystems (Figure 1.3). In the first case they aim at studying the dynamics of populations. In the second case they usually describe the flows of materials or energy among various parts of the ecosystem. Models may be either biological (i.e. simplified systems studied experimentally in the laboratory) or mathematical (investigation is carried out by means of analytical mathematical methods or simulations). Mathematical models may be governed by total causality, in which case they are known as deterministic, or may follow statistical distributions, in which case they are known as stochastic.

A mathematical model of an ecosystem is a mathematical description of some of its basic or all of its functions. In any case, a set of variables that describe the state of its various components must be distinguished. Such state variables of the ecosystem may be certain population sizes or trophic-level biomasses or accumulations of various chemical elements, etc. Certain other parameters, which also change over time, measure the external impacts

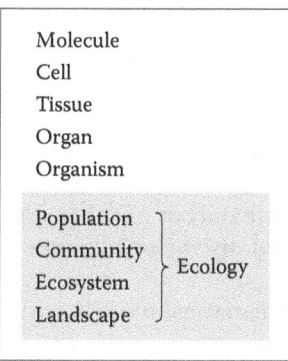

Figure 1.3 Levels of biological organization.

exerted on the ecosystem. Examples of these parameters, called *external factors*, are temperature, incoming solar energy, and trophic inflows.

A mathematical model of an ecosystem consists of functions that mainly connect the external factors and the state variables. These functions result, firstly, from the application of fundamental laws governing the physical, chemical, and biological phenomena of the functioning of the ecosystem.

The process during which sets of values of external factors are determined and, based on the functions of the model, the values of the state variables are calculated over time is called a mathematical simulation of the functioning of the ecosystem.

Usually, when studying ecosystems, the set of values of external factors in general is called *input*, and the resultant set of values of the state variables is called *output*.

The process of constructing a model generally includes the following stages:

- Determination of objective
- Selection of the necessary parameters for the given level of description, given that some information will be lost when moving up to each higher level
- Simplification of the system by omitting unimportant elements
- Formulation of hypotheses to facilitate the study
- Extraction and control of results

Present-day problems of protecting the environment have placed particular emphasis on ecology as an applied science. Mathematical simulation is a useful tool of ecology, in the effort to solve problems in relation to human management of ecosystems. As a rule, however, the results of a mathematical simulation do not lead to full, precise solutions to problems, for the following reasons, inter alia:

- Many ecological phenomena, particularly phenomena with a long-term evolution, have not been sufficiently investigated, and quantitative approaches to them are usually inadequate and/or superficial.
- Ecosystems are characterised by a complex interconnection, not only internal but usually also external. Thus, the separation of the biosphere into ecosystems is artificial, and therefore the realism of the ecosystem models is reduced. The input of the ecosystem includes elements that constitute output of other ecosystems, which influence, and are also influenced, by the one being examined.

Therefore, although mathematical models are a very useful tool for study, they are usually approximate, even rough, depictions of the functioning of natural ecosystems.

Chapter 2

Organisms and Environmental Factors

DOI: 10.1201/b14609-2

2.1 CATEGORIES OF ORGANISMS

Every living organism is an open system that exchanges matter and energy with the environment. In various ways the organism takes all the chemical substances it needs for the composition of its body from the environment and casts off into the environment the substances that are no longer useful for its functions. In addition, the organism takes from the environment energy in appropriate forms, which it uses to carry out the chemical reactions of synthesis, to move about, and to regulate its temperature; it casts off energy into the environment, mainly in the form of heat.

The sum of the physical and chemical processes occurring inside an organism, through which cellular matter is composed and decomposed, is called *metabolism*. The matter taken in by the organism and used to make cellular material or to produce energy is known as food. The processes of metabolism constitute basic links connecting the living organism with its environment and are fundamental elements of the functioning of ecosystems.

A feature of organisms is that they are composed of one or more cells, which are either similar to or different from each other and perform different functions. There is, however, another type of living things, the viruses, which are not composed of cells. Viruses are elementary forms of life; they cannot independently carry out the basic biological functions but are capable only of attacking the cells of various organisms and exploiting cellular functions in order to reproduce.

Organisms are classed in categories using various criteria such as:

- The complexity of biological organisation
- The source of carbon, which is the most basic component of organic matter and divides them into autotrophs and heterotrophs
- The source of energy, which divides them into photosynthetic and chemosynthetic organisms

Table 2.1 Combined Ranking of Organisms

Source of Carbon	Source of Energy	
	Photosynthetic	*Chemosynthetic*
Autotrophic	Plants	A few bacteria
	Most algae	
	A few bacteria	
Heterotrophic	A few bacteria	Animals
	A few algae	Protozoans
		Fungi
		Most bacteria

Source: Hadjibiros, K. (2007). *Ecology. Ecosystems and Environmental Protection,* 3rd edition. Symmetria, Athens (in Greek). With permission.

All the single-celled organisms, along with organisms composed of more than one cell but without differentiated functions, are characterised by a small degree of complexity. Composed of small, relatively incomplete cells, bacteria are the most elementary independent organisms. Fungi are many-celled organisms, whereas algae and protozoa may be single-celled or many-celled, but they are larger and more organised than bacteria. From the aspect of function, algae may be characterised as primitive plants and protozoa as primitive animals.

Plants and animals are characterised by differentiation of their cells and a significant degree of organisation and complexity. They are divided into the following categories (Table 2.1):

- Autotrophic organisms have the ability to synthesize organic from inorganic matter. In particular, they get the carbon they need from the CO_2 present in the atmosphere or dissolved in water.
- Heterotrophic organisms get the carbon they need from organic matter.
- Photosynthetic organisms take the energy they require from sunlight, through the process of photosynthesis.
- Chemosynthetic organisms take the energy they need from nutrients, by oxidizing organic or inorganic compounds.

2.2 CHEMICAL COMPOSITION OF THE CELL

The chemical elements C, H, and O are found in all organic compounds, whereas N is present in many of them. Seven other elements, P, S, Na, Mg, Cl, Ca and K, are present in life processes in significant but smaller proportions. There are at least another 25 chemical elements that are present in very small amounts in biological functions, as trace elements. The most important are: F, Si, V, Cr, Mn, Fe, Co, Cu, Zn, Se, Mo, Sn, I; an empirical formula gives the average occurrence of the six main elements in the composition of living matter: $H_{2960}O_{1480}C_{1480}N_{16}P_{1.8}S$. Significant deviations from this formula have been observed in the organic matter of the various categories of organisms.

The organic substances formed from the above-mentioned chemical elements may be divided into four main classes of polymeric compounds: polysaccharides, lipids, proteins and nucleic acids.

Polysaccharides are polymeric carbohydrates having the formula $(CH_2O)n$ with $n > 3$. They hydrolyse with ease or difficulty into simple carbohydrates (monosaccharides). Starch and cellulose are basic polysaccharides. Starch is easily hydrolysed and is used as food by many organisms. Cellulose is the most abundant organic compound on the planet, but most organisms, including humans, cannot hydrolyse it. Certain organisms do have this ability, among them cows and termites, which have the right sort of bacteria in their peptic system. Wood and cotton are examples of materials that are rich in cellulose.

Lipids include fats, oils and in general organic compounds that are soluble in organic solvents (e.g. benzine or ether) but are virtually insoluble in water. Fats and oils consist of C, H, and O, are more or less easily hydrolysed in organic acids and glycerine, and their biological role consists of storing combustible material. The lipids also include the lipoids, which may contain P or N.

Proteins are polymeric compounds, usually of great complexity, with high molecular weight (even over 1,000,000). They contain C, O, H and N in average proportions by weight: 50%, 23%, 7% and 16% respectively. S is also present in almost all the proteins, in a proportion of up to 3%. Proteins take part in most biological functions, e.g. as catalysts (enzymes) in biochemical reactions, as structural elements of cells and as transporters of matter (e.g. transport of oxygen to the cells by haemoglobin).

The nucleic acids are extremely complex compounds composed of hundreds to thousands of nucleotides. Depending on the carbohydrate that constitutes their base, we can divide them into DNA (deoxyribonucleic acid) and RNA (ribonucleic acid). They are spiral shaped, forming a double or single helix. DNA plays a crucial biological role, since it contains the chromosomes with genetic information on the structure and functioning of every organism. When a cell divides, each daughter cell has a full copy of the DNA of its parent cell, where all the hereditary information is recorded. RNA transports the information stored in the DNA to various parts of the cell, and in some organisms it is also involved in preserving genetic information.

2.3 METABOLISM—ENZYMES

The chemical reactions that make up metabolism are carried out with the help of enzymes. Enzymes speed up or catalyse reactions, without themselves undergoing permanent chemical change. The following is a simplified form of an enzymatic reaction:

$$F + E \rightarrow FE \rightarrow P + E$$

where F is food, P the final product, and E the enzyme. In reality, however, a complete enzymatic reaction consists of a series of secondary reactions, with a special enzyme intervening in each of them.

Enzymes are complex proteins. The maximum potency of an enzyme usually develops in a narrow zone, around a pH (usually pH = 7) and temperature value. Outside these zones, potency is very rapidly reduced, whereas above a certain temperature threshold the enzyme is destroyed.

The speed of an enzymatic reaction depends on the concentration of the food C_F and the concentration of the enzyme C_E. In general, the Michaelis-Menten relationship holds for elementary enzymatic reactions.

$$\frac{dC_F}{dt} = -\frac{KC_F}{K_S + C_F} - C_E$$

where time t; K, K_s are constants.

It is noted that if $C_F \ll K_s$ (low concentrations of food), the speed at which food is used is essentially proportional to its existing concentration. If $C_F \gg K_s$, speed is independent of the concentration of food.

2.4 STORAGE OF ENERGY

The energy obtained by organisms is temporarily stored in the cells, which then use it for their functions. The most basic of those functions is the endothermic synthesis of complex organic compounds from simpler organic or inorganic compounds. The energy produced by the energy reactions is transported in the form of chemical energy successively to the various organic compounds, generally with small thermal losses.

It is striking that, despite their differences, all cells use basically the same carrier of chemical energy. It is the nucleotide adenosine, which binds to phosphate radicals PO_4^{3-} and forms adenosine diphosphate (ADP) and adenosine triphosphate (ATP). The addition of one PO_4^{3-} to ADP is endothermic, resulting in the more energy-rich ATP. The conversion ADP $\rightarrow$ ATP entails storage, whereas the conversion ATP $\rightarrow$ ADP entails use of energy by the cell.

2.5 PHOTOSYNTHESIS—RESPIRATION— CHEMOSYNTHESIS

During photosynthesis, solar energy is converted into chemical energy through the excitation of the electrons of chlorophyll by the photons of solar radiation. The chemical energy is stored via the ADP $\rightarrow$ ATP reaction. At the same time, the photons cause photolysis of the water, again via chlorophyll. The breakdown of H_2O gives O_2, which is freed into the environment, while the H_2 is used to synthesize organic compounds. The reactions of the first stage of photosynthesis result in the manufacture of glucose, according to the overall equation:

$$6 \ CO_2 + 6 \ H_2O + \text{light energy} \rightarrow C_6H_{12}O_6 + 6 \ O_2$$

Then the glucose molecules bind to nitrogen and other substances, resulting in the complex organic compounds. The reactions are endothermic, and the energy required is provided through the breakdown of ATP into ADP.

During chemosynthesis, the cells take the energy required through oxidation of organic or inorganic substances, via the process of respiration.

Oxidation is the removal of electrons, whereas reduction is the addition of electrons. When an organic material is oxidized biochemically, it loses electrons, generally in the form of hydrogen atoms. The H that is removed is transported to the final recipient by a series of enzymes. In the case where the final recipient of the H atoms is O_2 (of the air or water), H_2O is formed and the oxidation is called *aerobic*, whereas in all other cases it is called *anaerobic*. The chemical energy produced is stored via the ADT $\rightarrow$ ATP reaction.

Aerobic respiration is shown in the overall equation:

$$\text{Organic matter} + O_2 \rightarrow CO_2 + H_2O + \text{heat}$$

All plants and animals and most of the other organisms draw energy through the process of aerobic respiration. There are, however, certain categories of bacteria that carry out anaerobic respiration: as final recipients of the H atoms, they use $SO_4^=$ in producing H_2S, NO_3^- in producing N_2, CO_2 in producing CH_4, etc. These bacteria are either strictly anaerobic, that is, they appear only in conditions where O_2 is absent, or facultative, that is, aerobic in the presence of O_2 and anaerobic in the absence of O_2.

During reduction of CO_2, methane CH_4 is produced. An odourless, non-toxic gas, methane is a well-known product of septic (anaerobic) processes. When mixed with atmospheric air in various proportions, methane gives an explosive gaseous mixture, which may cause serious accidents in waste cleaning works and sewers. Particularly well known because of its strong unpleasant odour, and also very toxic, is hydrogen sulphide H_2S.

It is often emitted by eutrophic water bodies, septic tanks and sewers where the oxygen dissolved in the waste has been exhausted, initially by the action of aerobic microorganisms. Later, anaerobic microorganisms develop, some of which use the sulphate present in waste as a recipient of hydrogen, resulting in the production of H_2S.

Autotrophic organisms generally supply heterotrophic organisms with nutrients consisting of various complex compounds, such as polysaccharides, proteins and fats. These substances cannot pass through the cell walls and must first be broken down into simpler ones, such as carbohydrates, amino acids and fatty acids. Most microorganisms break them down as necessary using special extracellular enzymes, whereas higher organisms use their specialized peptic system. In any case the heterotrophic cell ultimately synthesizes its complex components from the simpler foods that enter the cell and from the products of their breakdown, which is carried out within it. The reactions of synthesis are endothermic and the necessary energy is provided by ATP. In the absence of sunlight and food, photosynthetic and chemosynthetic cells may obtain the energy they need through oxidation of their protoplasm. This phenomenon is known as endogenous respiration.

The reactions of synthesis are interconnected with the reactions of energy and form a total of at least 1000 secondary chemical reactions, which are necessary to support the life of the cell. Observation of the small, single-celled organism *Escherichia coli* (coliform bacterium), for example, shows it to be an astonishing chemical factory. Within that microscopic cell an average of 1400 molecules are produced every second. As a comparison, in humans' initial efforts to produce a protein, successive chemical reactions lasting for several months are necessary.

2.6 LIMITING NUTRIENT

An organism requires chemical elements in proportions that are in general determined by the chemical composition of its cells, and thus it uses available nutrients in those proportions. Thus the conclusion can be drawn that the quantity of living matter produced at a certain moment in time in given environmental conditions will be limited by the chemical element that is least available in those proportions, barring the influence of other negative factors. This is a special formulation of von Liebig's "law of the minimum" (Chapter 4).

In general, the limiting factor is some environmental parameter. Often it is one component of the nutrients, in which case it is called the *limiting nutrient*. For instance, in terms of mass, on average C, N, P are present in the protoplasm of algae in the proportion of C:N:P = 41:7:1. Assuming that there is a surplus of other elements (oxygen, hydrogen, trace elements, etc.), energy, and the other factors that play some part in production, the limiting nutrient will be whichever one of C, N, P is least available with regard to the above proportion. Decreasing the limiting nutrient entails a proportionate decrease in the valorization of the other nutrients by the organism and an approximately proportionate decrease in the biomass produced.

2.7 INTERACTION OF ORGANISMS AND ENVIRONMENT

Every living organism interacts with its environment in a variety of ways. The various characteristics that enter into the study of this interaction are called *environmental factors*; they have traditionally been divided into physical factors such as radiation, the wind and the temperature of the air or water, chemical factors such as concentration of O_2, CO_2 or nutrient salts, and biological factors such as the presence of organisms of the same or another species. The physical and chemical factors create the abiotic environment, and the biological factors create the biotic environment.

The biotic environment is examined in population studies (Chapter 3), the chemical factors in the environment are examined in the context of biogeochemical cycles (Chapter 4) and pollution (Chapters 7, 8, 9), and the physical factors are examined next, also in the context of energy flow (Chapter 4).

2.7.1 Physical Factors

Organisms are affected by many physical factors in the environment, which often interact or act in a complex manner. Thus an organism's response to a physical factor may vary, depending on the effect of other factors. For example, plant transpiration is reduced if humidity is high, and increases with the wind if humidity is low. Isolation of the relationship between each factor and the response of the organism may give rise to misleading conclusions. Nevertheless, a breakdown into individual factors is often useful and a greater or lesser breakdown of the study of the relevant ecological phenomena is in general unavoidable, but particular attention must be paid to the drawing of conclusions. On the other hand, the viewpoint that the environment can be studied only as a unified whole often proves to be excessive and fruitless.

The basic physical factors of the environment (i.e. temperature, wind, humidity and radiation) have a common basis, which makes it possible to study their combined effects on the organism being examined. Specifically, these factors entail a transfer of energy to and from the organism, through processes of evaporation, absorption or emission of radiation, and conduction or convection of heat. Regarding energy changes, the organism's thermal balance equation can be written, which links the physical factors and assists the study of their combined effect on the organism. Its general form is the following:

$$R_s + R_e + M = R_0 \pm H + E,$$

where:

R_s, R_e = increase in the organism's thermal energy due to absorption of solar and other radiation respectively

M = thermal contribution of metabolism

R_0 = energy emitted by the organism in the form of radiation

H = exchange of energy between the organism and the environment due to conduction and convection (or transfer) of heat

E = loss of heat from the organism due to evaporation

Next we give a brief qualitative description of the physical factors of the environment.

2.7.1.1 Temperature, Wind, Humidity

Temperature is the physical factor that has been measured the most. It is a main regulatory factor in heat exchange between the environment and the organism.

The following observations have been made with regard to an organism's response to temperature. Heat conduction takes place through molecular motion, without transfer of matter. The molecules maintain their relative positions and transmit only momentum. A classic example is the flow of heat from the hotter to the colder end of a metal rod. By contrast, heat convection (sometimes called *heat transfer*) requires the existence of a fluid medium and takes place through the motion of elements of the fluid. During the usual turbulent flow of atmospheric air or water, the main heat transfer mechanism is convection. For an organism exposed to the atmosphere, the speed of heat convection is proportional to the difference in temperature between the organism and the air and is also proportional to wind velocity.

For every organism there is a temperature range within which it can live. For plants a common maximum is 45°C and a common minimum –10°C. However, there are many plants that can live at temperatures outside this range. Photosynthesis generally accelerates when the temperature increases, up to a limit, beyond which it slows down.

Depending on their response to the temperature of the environment, animals are divided into homeotherms and poikilotherms. Homeothermic animals (i.e. birds and mammals) produce enough thermal energy through the process of metabolism to keep their temperature relatively stable, regardless of the temperature of the environment. The lower the temperature of the environment T_e in relation to the temperature of the organism T_o, the stronger metabolism has to be. At environmental temperatures of $T_e > T_o$, the organism gives off heat through the evaporation of water. Activation of water for evaporation also requires consumption of energy.

Poikilothermic animals (fish, reptiles, etc.) are able to exercise only slight internal control over their temperature. The temperature of the environment largely controls the rates of their biochemical functions, which approximately double for every 10°C temperature increase. Thus, these organisms regulate their temperature mainly through the appropriate behaviour. Fish choose the area of the aquatic environment that is at the right temperature. Reptiles regulate their temperature by moving between shade and sun, by changing their orientation in relation to the sun, and even by changing their shape.

It is noteworthy that the death toll inflicted on fish and other aquatic organisms by thermal pollution of bodies of water (Chapter 8) is associated with the poikilothermia of those organisms. Thermal pollution is a rise in water temperature due to drainage of warm water, which has usually been used by power plants and other factories for cooling purposes. The high mortality rate is due to suffocation, because it causes the fishes' metabolism to rise and thus their requirements for oxygen increase proportionally, to such an extent that the oxygen dissolved in the water is not enough. In addition, the higher water temperature causes a decrease in the concentration of dissolved oxygen.

The wind causes air masses in contact with an organism's body to be replaced by other masses that may contain different amounts of heat, water vapour or carbon dioxide. Thus the wind affects heat exchange between the organism and the environment, as well as the elimination of water and carbon dioxide by the organism.

Atmospheric humidity is of importance mainly due to its impact on the rate of evaporation from the bodies of organisms.

2.7.1.2 Radiation

The radiation an organism receives is terrestrial or extra-terrestrial in origin. In the first case it originates from various bodies on Earth. In the second case it originates from the sun or is cosmic radiation. Solar radiation is of primary importance for the functions of ecosystems, because for them it constitutes a basic source of energy (Chapter 4). It also plays a decisive part in the planet's thermal equilibrium and affects organisms' survival (Chapter 9).

Depending on its wavelength (in $\mu = 10^{-6}$ meter) or its frequency, the spectrum of electromagnetic radiation can be divided into the following regions:

- Microwaves and radio waves (wavelength >1000 μ)
- Infrared radiation (wavelength 0.7–1000 μ)
- Visible light (wavelength 0.4–0.7 μ)

- Ultraviolet radiation (wavelength <0.4 μ)
- Roentgen or x-rays (wavelength <10^{-2} μ)
- Gamma rays (wavelength <10^{-4} μ), which belong to the category of ionising radiation, together with alpha and beta radioactivity (Chapter 7)

When radiation falls on a body, it can continue on its path in three possible ways: by reflection, transmission or absorption (Table 2.2; Figure 2.1). Each of these phenomena can occur for all or part of the incident radiation. Transmission through the body occurs when the body is transparent to the specific radiation, as is water to visible light or a wall to gamma radioactivity, for example. Reflection is associated with the form of the surface and the material of the body: for example, light is reflected by the surface of a lake or by snow. Absorption entails conversion of radiation energy into thermal or another type of energy or into a different type of radiation (e.g. a stone absorbs part of incident visible light, heats up, and emits infrared radiation). The absorption of a small proportion of sunlight (about 1%) by plants and algae sustains the phenomenon of photosynthesis and the energy flow in the ecosystems; it constitutes the most significant result of the action of solar radiation on organisms.

Of the solar energy that reaches the outer limits of the atmosphere, around 9% lies in the ultraviolet, 41% in the visible and 50% in the infrared portion of the spectrum (Figure 2.2). A substantial part of this radiation is absorbed by atmospheric gases and by suspended solid and liquid particles (clouds, smoke, dust, suspended salts, etc.). Another part is reflected and ultimately returns to space.

Most of the ultraviolet portion of incident radiation is absorbed by the atmospheric gases O_3, N_2, and O_2, and a substantial part of the infrared portion is absorbed by H_2O and CO_2. With the exception of clouds, which

Table 2.2 Transformation of Solar Energy as Percentage of Annual Input into the Biosphere, the Atmosphere, and the Hydrosphere

Energy Transformation	Percentage
Reflection	30
Direct conversion into heat	46
Evaporation, precipitation	23
Wind, waves, currents	0.2
Photosynthesis	0.8
Total	100

Source: Modified from Odum, E.P. (1997). *Ecology. A Bridge between Science and Society.* Sinauer, Sunderland, MA; and from Hadjibiros, K. (2007). *Ecology. Ecosystems and Environmental Protection,* 3rd edition. Symmetria, Athens (in Greek). With permission.

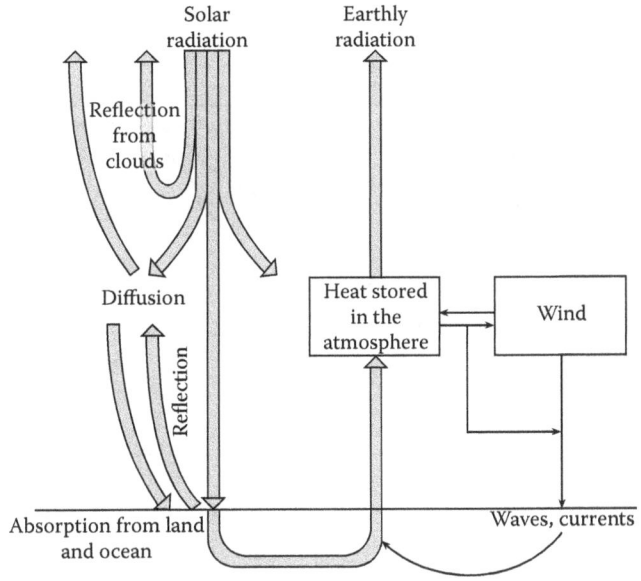

Figure 2.1 Conversions of incident solar radiation. (Modified from Hadjibiros, K. (2007). *Ecology. Ecosystems and Environmental Protection,* 3rd edition. Symmetria, Athens (in Greek). With permission.)

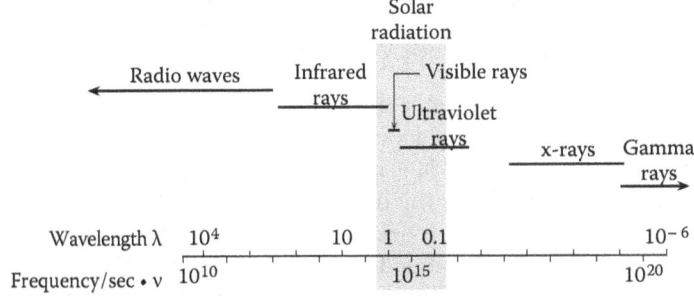

Figure 2.2 The spectrum of electromagnetic radiation. Solar radiation is in the middle range of the spectrum. (From Odum, E.P. (1997). *Ecology. A Bridge between Science and Society.* Sinauer, Sunderland, MA. With permission.)

reflect part of the light, the atmosphere is in general transparent for the visible portion, most of which reaches the surface of the Earth in the form of direct and diffuse solar radiation. The surfaces of the land and the sea reflect part of the radiation back into space, according to the reflectivity coefficient (albedo), which depends on the nature of the surface. This reflected portion can be a small, medium or large proportion of the incident radiation, as the case may be. On forested or aqueous surfaces it does

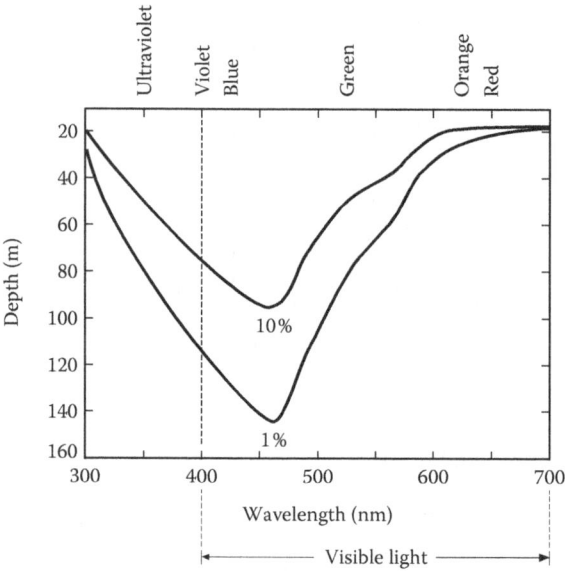

Figure 2.3 Light extinction by clear water. (Jerlov, N.G. 1951. Optical studies of ocean water. At depths at which 10% and 1% of normal sunlight penetrate into the clearest seawater, red light is absorbed rapidly, but even the blue end of the visible spectrum is mostly extinguished by 140 m. The deeper a plant lives, the more it is dependent on a narrow band of wavelengths. *Reports of the Swedish Deep Sea Expedition,* 3: 1–59, modified by Colinvaux, P. 1993. *Ecology* 2. Wiley, New York. With kind permission from John Wiley & Sons.)

not exceed 3–5%. On the surface of freshly fallen snow it can be as high as 95%. Reflection is also greatly dependent on the angle of incidence. That is, when the sun is at its zenith, 5% is reflected by an aqueous surface, whereas when the sun is low (5° above the horizon) 95% is reflected.

Part of solar radiation is reflected from the surface of the water or from bodies floating or suspended at shallow depths. The rest enters the body of water but is gradually absorbed by the H_2O molecules, aquatic organisms and dissolved or suspended substances; as a result its intensity rapidly decreases with depth. Part of the light is absorbed by algae through photosynthesis. When the water is turbid (e.g. in lakes, coastal areas), the decrease is much greater. Even in clear water, sunlight is greatly weakened at a depth of a few dozen meters (Figure 2.3). Based on its illumination, a body of water can be divided into the following zones:

- Euphotic zone, where photosynthesis produces more oxygen than total respiration consumes (Chapter 8)
- Dysphotic zone, where photosynthesis produces less oxygen than total respiration consumes (Chapter 8)

- Aphotic zone, which no light reaches and where no photosynthesis takes place

Radiation decreases exponentially in water, that is, the correlation of strength at the surface I_0 and strength I_z at a depth z is given by the relationship:

$$I_z = I_0 e^{-kz}$$

where z = depth below the surface (Chapter 8).

In temperate oceans $k = 0{,}10 - 0{,}20$ m^{-1} has been measured, whereas in coastal waters, the coefficient k can be as high as $k = 1{,}0$ m^{-1}. In very turbid lakes values for k as high as 3–4 m^{-1} have been measured.

Organisms, like all bodies for that matter, have a specific distribution of the spectrum of absorbed visible and infrared radiation. Thus, various organisms exposed to a given radiation can absorb different amounts of energy. On the other hand, all the bodies on the surface of the Earth, including organisms, radiate energy in the infrared part of the spectrum, at wavelengths of 9–11 µ. The intensity of the emitted radiation depends on surface temperature.

According to the Stefan-Boltzmann relationship, this is:

$$R_0 = \varepsilon \sigma T^4 \, \mathrm{cal} / \left(\mathrm{cm}^2 / \mathrm{sec} \right),$$

where:

   T = temperature in Kelvin
   σ = Stefan-Boltzmann's constant
   ε = ratio of the actual radiation to the radiation of a perfect radiator/absorber (black body), at the same temperature

For almost all bodies of ecological importance, such as plant leaves, animals, etc. ε ranges between 0.95 and 0.98.

The Stefan-Boltzmann relationship explains the manner in which the planet's thermal equilibrium is achieved. Earth receives huge amounts of solar energy capable of causing a constant increase in its temperature. This does not occur, however, because the rise in temperature very rapidly increases (in proportion to T^4) the emission of energy in the form of infrared radiation, ultimately resulting in the maintenance of the thermal equilibrium. Thus, on average the planet's surface temperature remains constant. The greenhouse effect (Chapter 9) tends to raise the temperature at which thermal equilibrium occurs.

Chapter 3

Organisation at Population Level

DOI: 10.1201/b14609-3

3.1 POPULATION AND ECOSYSTEM

The study of the living elements of an ecosystem could theoretically mean the examination of the actions of all organisms, which are extremely numerous. Therefore, for practical reasons, organisms should be grouped, and individuals of the same species that live in the same space should be examined as a whole. Thus the concept of population results, which constitutes a basic ecological unit. A significant part of the phenomena observed at a population or ecosystem level can be explained through the interactions of populations.

Populations are formed by groups of individuals of the same species that live in a certain place so that they can breed with one another. The total of populations living in one place is called *biological community*. Important issues concerning population ecology are the prediction of:

- Population change over time
- Population distribution in space
- The abundance of each population within the biological community
- The distribution of the individuals of a population by age and sex

Population is a characteristic case of a system. Even though population is nothing more than the total of individuals comprising it, it can be described by parameters and variables, such as population size, natality, mortality, immigration, age distribution etc. that have no meaning for each one of the individuals but only for the population itself. These parameters determine in any given moment the state of the population, its temporal change and the manner in which external circumstances affect it.

The ecological niche is an important concept of theoretical ecology. It represents the space of a population's ecological activity, i.e. the natural space where it lives, its trophic and other relationships with the rest of the organisms, its place in relation to the environmental grading of the physicochemical factors, and the modifications it causes to the environmental conditions. Generally, in an ecosystem there is a one-to-one correspondence between the total of the populations and the total of available niches. The niche is not a geographical concept but may be considered as a multidimensional super-space, where each dimension represents an important ecological parameter and within which the environment allows the survival of the population. The preservation of a species is connected with securing its populations' ecological niche in the biosphere.

A sub-total of the concept "ecological niche" is the ecological habitat, i.e. the set of the geographical, abiotic and biotic conditions of the natural environment that suit a population. In the past, the biotope had a similar meaning. Currently, the term *biotope* is mostly used to define a geographical area important for the conservation of the natural environment because one or more protected biological species dwell there. The adoption of the biotope concept gives an existing and expanding use of the term.

The growth of a population is affected both by its internal forces and external factors. As a rule, the population has an internal tendency to increase with a fixed rate per individual, which initially leads to an exponential change. However, the rate of increase is usually halted under the influence of abiotic environmental factors such as temperature, radiation, humidity, nutrients etc. or biotic environmental factors such as simple or complex population interactions within the ecosystem. One of these factors emerges in each case as the limiting factor.

A population's size change in time basically depends on three parameters:

- Natality rate
- Mortality rate
- Immigration rates of individuals to and from the population

The examination of temporal change within animal and plant populations, as well as that of the human population, has always been of vast practical importance. That is why many mathematical methods of study, mainly of empirical/statistical character, were developed a long time ago in order to assess these phenomena; these methods belong to population dynamics and demography.

3.1.1 Age Distribution

Each population can be subdivided into age groups. All the individuals of a population whose age lies within a specific age span belong to the corresponding group. For example, from 100 individuals of a human population, 20 may be children (0–15 years), 15 may be youngsters (15–25), 40 mature (25–60) and 25 elderly (over 60) (Figures 3.1, 3.2 and 3.3). Characteristic age distributions of human populations are shown in Figure 3.4.

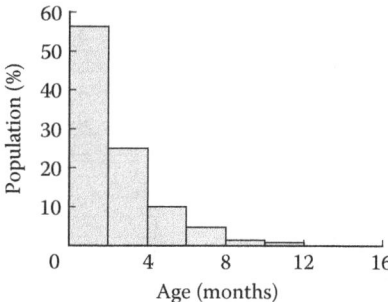

Figure 3.1 Age distribution of a population with high natality rate. (From Hadjibiros 2007. *Ecology. Ecosystems and Environmental Protection,* 3rd edition. Symmetria, Athens. With permission.)

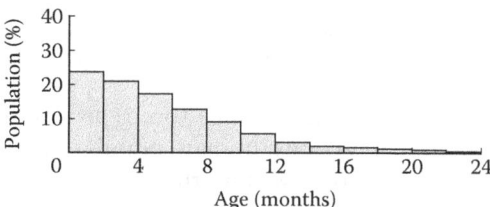

Figure 3.2 Age distribution of a stagnant population. (From Hadjibiros 2007. *Ecology. Ecosystems and Environmental Protection,* 3rd edition. Symmetria, Athens. With permission.)

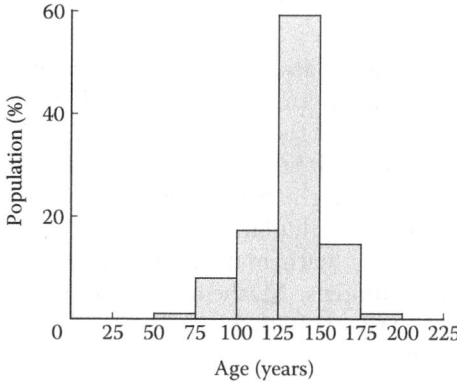

Figure 3.3 Age distribution of a tree non-generating population in a mountain forest. (From Hadjibiros 2007. *Ecology. Ecosystems and Environmental Protection,* 3rd edition. Symmetria, Athens. With permission.)

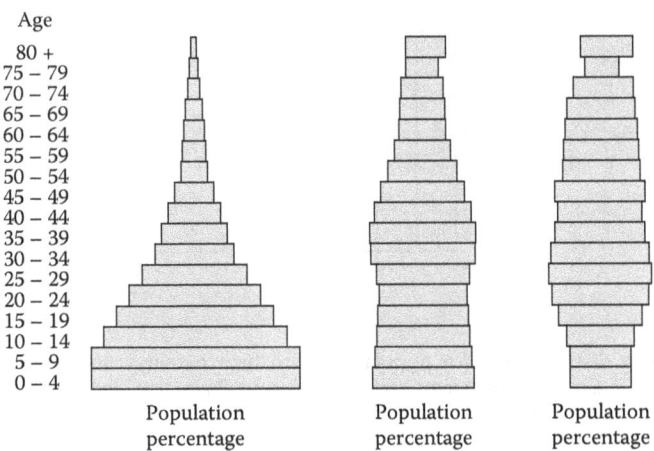

Figure 3.4 Pyramids of three human populations, with increase rates 2.1%, 0.6%, and 0.0% per year. (From Hadjibiros 2007. *Ecology. Ecosystems and Environmental Protection,* 3rd edition. Symmetria, Athens. With permission.)

Examining the three basic age groups (pre-reproductive, reproductive, and post-reproductive) in various human populations, we can observe characteristic age distributions within the relatively stabilized populations of developed countries compared to the certain developing countries' populations. The biometrician Lotka demonstrated that if natality and mortality are temporally invariable, then after a long enough time period, the population's age distribution also becomes stable.

3.2 POPULATION DYNAMICS

The study of diachronic population variations constitutes a branch of ecology known as population dynamics. Its purpose is to examine the way in which population sizes change through time, along with the analysis of the causes of these variations. This study is of great practical importance given that humans are interested in the prediction, or even control, of the size of various human or other animal, plant or micro-organism populations.

Population dynamics is a mathematical branch of ecology that uses various simple or complex models. Mathematical analysis of human populations is similar to the analysis of other animal populations with overlapping generations. In population dynamics, deterministic or stochastic models, differential equations, or other mathematical tools such as matrices are used. In simpler models, a basic assumption stating that all population individuals are identical is followed; therefore the population state can be described with only one variable—its size, N.

A central issue for ecology is the search for the causes of ecosystem stability, which are considered to be connected with the interactions between
populations. The explanation of ecosystem stability represents a main theoretical and practical challenge for population dynamics. The existence of
stability, a precondition for persistence, is an important criterion for evaluating models used to describe ecological phenomena. However, the high
complexity of ecosystems does not allow, at least for the time being, the
formulation of a stability theory. Consequently, all theoretical approaches
are limited to the exploration of stabilising factors which could eventually
contribute, to a lesser or greater degree, to the observed stability of the
ecosystem. Such factors might be intra- or inter-population competition,
predation, the diversity of species or of interactions or of the natural conditions, spatial or temporal lack of homogeneity that creates possibilities
for the diminishing population to seek refuge, the natural change of the
parameters due to populations' evolution etc. Other factors, such as the
anthropogenic modifications, could hardly be stabilizing.

3.3 POPULATION SIZE VARIATION MODELS

3.3.1 Exponential Model

If we accept the preconditions mentioned below, the population variation
rate per individual will remain constant, i.e. each individual will produce
a fixed number of offspring in a given time period. Thus, the following
simple model results (Figure 3.5):

$$\frac{dN}{dt} = rN$$

where dN/dt is the population variation rate. The constant variation rate
per person, r, is equal to the difference natality rate minus mortality rate
of the population, and is defined as the intrinsic rate of natural increase.

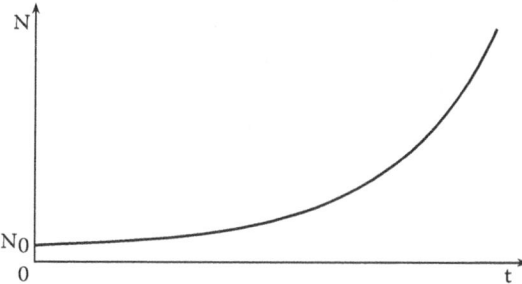

Figure 3.5 Exponential increase. (From Hadjibiros 2007. *Ecology. Ecosystems and
Environmental Protection,* 3rd edition. Symmetria, Athens. With permission.)

The result of the above equation's integration is:

$$N = N_o e^{rt}$$

where N_0 is the initial population size (i.e. when $t = 0$).

This exponential model describes an unlimited growth of the population. Its two basic preconditions are that the population inhabits an unlimited space and that unlimited resources are available for its development, i.e. its growth is not inhibited by any limiting factor (Chapter 4, Section 4.2). These conditions are not usually observed in nature. However, they are possible in some cases for some time period; in that case, an exponentially increasing population may be observed.

3.3.2 Logistic Model

Each population has the natural ability to increase exponentially, but its growth is inhibited by the biotic and abiotic environmental influences. The available resources are not inexhaustible in nature, and sooner or later a limiting factor will appear. Thus population increase will intensify competition between its individuals who are trying to claim the same resources, and in particular the resource represented by the limiting factor. This kind of competition is defined as intraspecific; by contrast, interspecific competition is that between different populations. Under these circumstances of intraspecific competition, the population exhibits self-limitation and its temporal variation can be described by an equation of the following type:

$$\frac{dN}{dt} = rN - qN^2$$

where q is a constant. In this case, the variation rate per person is not stable. The internal natural increase rate, r, is a constant but not equal to natality minus mortality. This equation defines the logistic model of population increase, which in general is graphically represented by a sigmoid curve (Figure 3.6). The initial part of the curve corresponds, approximately, to an exponential increase. After the point of inflection, the rate of increase is significantly decelerated and is marginally nullified due to high intraspecific competition.

Setting $q = (r/K)$ in the initial equation, we have:

$$\frac{dN}{dt} = rN\left(1 - \frac{N}{K}\right)$$

We observe that:

- For $N \ll K$, increase is practically exponential
- For $N \to K$, increase constantly decelerates

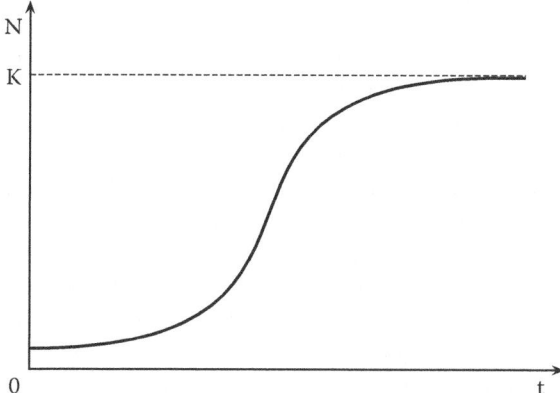

Figure 3.6 Logistic increase. (From Hadjibiros 2007. *Ecology. Ecosystems and Environmental Protection,* 3rd edition. Symmetria, Athens. With permission.)

- For $N = K$, the variation of population size becomes zero; the population is stabilized
- For $N > K$, the size of N decreases until $N = K$

Therefore, the value of K is the maximum size of the population which can survive in a given environment. It is called *carrying capacity* of the environment, and its value is defined every moment by the current limiting factor (Chapter 4, Section 4.2), because it is the manifestation of its action in the particular situation. Potential elimination of the limiting factor would lead to exponential increase and a possible population boom. In the case of a new limiting factor emergence, the system returns to a self-limiting increase but with a different carrying capacity. With the occurrence either of a change in natural conditions or an appropriate human interference over the course of time, a succession of limiting factors and corresponding carrying capacities will appear.

The integration of the logistic model (for stable K) gives:

$$N = \frac{K}{1 + \left[(K/N_0) - 1 \right] e^{-rt}}$$

The logistic model gives, in many cases, a satisfactory description of the population variations encountered in nature. Such an example is the increase of unicellular organisms with a small life span and great reproduction speed. However, in most cases of natural populations the variation is more complex and difficult to interpret. Nevertheless, the logistic model, as also the exponential one, has many applications in population dynamics.

3.4 INTERACTIONS BETWEEN TWO POPULATIONS

The possibilities of interactions among populations are plenty and so are their consequences, which can be classified according to the biological interactions that take place. Yet considering the astonishing diversity in the manner in which various plants and animals compete for space and food, support each other, seek for or serve as food, it becomes obvious that this classification is ineffective. Ultimately, each biological interaction constitutes a distinct situation. This difficulty can be overcome if these interactions are classified not simply on the basis of the processes involved, but on the quantitative results they bring about. This approach is the natural outcome of the populations' mathematical description and offers, besides simplicity, the advantage of its conclusions being translated into quantitative relationships, and therefore into mathematical models. The results of these interactions can be characterized as positive, negative or neutral, depending on whether a population size increases, decreases or is not affected by the other population's presence. For example, from the equation $dN_1/dt = \alpha_{11}N_1 + \alpha_{21}N_1N_2 - \beta_{31}N_1N_3$ we conclude that the change of population N_1 is subjected to a positive effect by N_2 and a negative one by N_3, while the possible presence of a N_4 population has a neutral effect.

If we accept the fact that the interactions between two populations quantitatively influence the change of their size, we can classify them according to the positive (+) or negative (–) or absence of impact by population A on B or vice versa. Most interesting are the cases:

++, defined as symbiosis
– –, defined as competition
+–, defined as predation or predator-prey interaction

It must be noted that the interaction among populations constitutes one of the main factors that shape natural selection; therefore it affects, to a certain degree, the evolution of the respective species.

3.4.1 Symbiosis of Two Populations

Symbiosis of two populations is an important biological function. Nevertheless, its effects on population and ecosystem dynamics are rather limited and have not been the object of extensive studies.

In a symbiotic relationship, the growth of each population is positively affected by the presence of another population. This interaction tends to lead the system of the two populations to a more rapid growth or to expand the carrying capacity of the environment. If one of the populations disappears, then the second one, if it is completely dependent on the first, will follow the same path. In the case of its survival, it will be increasing in a slower rate or will equilibrate in a smaller size.

3.4.2 Competition between Two Populations

The interaction of (interspecific) competition corresponds to the relationship between two populations that both need to exploit some of the limited resources present in their environment (space, light, nutrients etc.). The available resources are not inexhaustible, and sooner or later, a limiting factor will emerge. The increase of populations intensifies the competition between them, especially for the resource which is represented by the limiting factor. Each one of the populations attempts to inhibit the other's growth and be the only one that will benefit from the available resource.

Exploitation competition arises when the availability of a resource for each one of the populations is reduced due to the use of the resource by the other population. Interference competition is manifested by a special behaviour (e.g. territory defence) or by an influence of chemical form (e.g. through a toxic substance) towards the competitor in order to prevent his exploitation of the resource. Generally, a necessary precondition for the existence of competition is an important overlap of the two populations' ecological niches.

One of the competition's outcomes is the elimination of the weakest competitor. The competitive exclusion principle, a significant hypothesis of theoretical ecology known as Gause's law, states that when two populations claim the same limited resources, their competition cannot persist infinitely if the environmental conditions remain stable. One of the two populations will prove more capable in resource use and will dominate, whereas the other will perish.

This is a necessary outcome, assuming environmental conditions remain fixed so that the relative advantage possessed by one of the populations can last long enough, i.e. until its competitor is eliminated. If environmental conditions are changing so that they benefit both competing species in turn, it is possible that none of the populations will maintain its advantages long enough to eliminate its competitors; in this case, a permanent coexistence may result. An essential precondition for the final predominance of one of the competitors is for both species to demonstrate stable abilities. Competitive species which succeed in evolving so that they achieve adequate differentiation of ecological niches, may continue to coexist.

Apart from the temporal variability of the environmental factors, coexistence preconditions include spatial variability. In an area of the ecosystem, one of the competitors might have advantages against the other, causing the other's elimination, whereas the opposite can be occurring in a different area of the ecosystem, whereon it is possible for the competitors to coexist at the whole ecosystem level. Other potential coexistence factors can be predators (predator organisms or parasites) which, by attacking selectively the more abundant species, can prevent it from increasing enough to be able to eliminate its competitor.

3.4.2.1 Competition Models of Two Populations

Competition can be described by the typical Lotka-Volterra equations, according to the exponential model:

$$dN_1/dt = r_1 N_1 - \alpha_{21} N_1 N_2$$

$$dN_2/dt = r_2 N_2 - \alpha_{12} N_1 N_2$$

The corresponding equations according to the logistic model are:

$$dN_1/dt = r_1 N_1 - \alpha_{21} N_1 N_2 - q_1 N_1^2$$

$$dN_2/dt = r_2 N_2 - \alpha_{12} N_1 N_2 - q_2 N_2^2$$

In the first case, one population's rate of increase decreases due to the other population's presence. In the latter case, each individual from both populations is subjected to a double competition, interspecific and intraspecific. The result is decreased growth rates of both populations, and sizes which cannot reach the environmental carrying capacity that each population would achieve without the impact of the competition. However, if one of the populations disappears, the other will increase more rapidly, enjoying the whole of the environmental carrying capacity.

Coefficients α_{21} and α_{12} regulate the negative effect of one of the populations to the increase rate of the other; this effect depends on the value of the product $N_1 N_2$. The presence of this product is necessary so that each population gets rid of interspecific competition if its competitor disappears, whereas if its own size becomes zero, the rate of its change becomes zero, too.

The above equations cannot be solved analytically. In order to explore the system's equilibrium points, in the case of the logistic model, we write them as follows:

$$dN_1/dt = r_1 N_1 \left(1 - \left(N_1 + \alpha_{21}^* N_2 \right) / K_1 \right)$$

$$dN_2/dt = r_2 N_2 \left(1 - \left(N_2 + \alpha_{12}^* N_1 \right) / K_2 \right)$$

where $K_1 = r_1/q_1$, $K_2 = r_2/q_2$, $\alpha_{12}^* = \alpha_{12} K_1/r_1$, $\alpha_{21}^* = \alpha_{21} K_2/r_2$.

We observe that: $dN_1/dt = 0$, if $K_1 = N_1 + \alpha_{21}^* N_2$, and $dN_2/dt = 0$, if $K_2 = N_2 + \alpha_{12}^* N_1$. Placing the two lines, $K_1 = N_1 + \alpha_{21}^* N_2$ and $K_2 = N_2 + \alpha_{12}^* N_1$

in the phase space (two-dimensional space N_1 and N_2 where time is absent), and conducting the related mathematical exploration we find that:

- If $K_1 < r_2/\alpha_{12}$ and, at the same time, $K_2 < r_1/\alpha_{21}$, the system is in a stable equilibrium, thus both populations permanently coexist. This occurs because the size increase of one population inhibits its own growth more than it inhibits the growth of its competitor. In this case competition can last infinitely, because it has low enough intensity, i.e. Gause's law does not apply.
- In every case where the previous condition does not apply, the system is not in stable equilibrium, therefore the competition cannot continue for a long time. Depending on the parameters' values, one of the two populations will predominate while the other will be eliminated.

Even though the above analysis is based on a very simple competition model, its qualitative conclusions cover all possible outcomes of a two species competition. The power of the competitive exclusion principle becomes obvious, though it does not have a universal application. Research has shown that there are examples in nature of two species' coexistence when the intensity of competition is reduced for various reasons, such as the variability of the species' characteristics (e.g. through natural selection) or environmental variability (e.g. through selective predation, periodical fluctuation of natural factors etc.).

Simple competition models represent approximations of a strategic nature. Their essential qualitative conclusion is that competition cannot cause permanent oscillations to populations' sizes, but it may lead to the elimination of one of the competitors or, under specific preconditions, to coexistence of the two populations. Therefore, competition has the ability to lead to the alteration of an ecosystem's composition.

3.4.3 Predator-Prey Interaction

The most significant and widespread interaction between two populations is the predator-prey interaction. It corresponds to the general trophic relationship where one organism consumes the other, i.e. the interactions between carnivore and herbivore, herbivore and plant, parasite and host etc.

The term *predator* mostly refers to carnivores, although it can be generalised to include herbivores and parasites. Parasites are organisms that feed on their host, which is either not killed or is killed at a slow rate. Various funguses, bacteria and insects represent parasite examples. General predator-prey interaction can be divided into four categories:

- Predators that kill numerous preys (mostly carnivores and some herbivores)

- Predators that consume only a part of numerous preys without killing them (mostly herbivores)
- Predators that kill only one prey during their life span (mainly some insects that are parasites of other insects)
- Predators that consume a part of only one prey in their life span (most parasites)

Trophic relations do not always coincide with the notion of predation as a quantitative interaction between two populations. In nature, trophic relations between two populations can result in the decrease of one population and the increase of the other, but sometimes they may also be beneficial for both populations or have a neutral effect on one and positive on the other etc.

In nature, observed interactions present a great diversity due to the existing differences between species, and also due to numerous environmental factors. Generally, the predator-prey interaction does not result in the elimination of one or both populations. A typical example is the permanent oscillation of both populations with a phase difference (Figure 3.7). Prey decrease may be followed by predator decrease (due to lack of food), which, though, leads to an increase in prey numbers (due to decrease of predation), which in turn may cause an increase in predator numbers (due to excessive food), which brings about a decrease in prey numbers etc. As a rule, such a system exhibits a notable elasticity, but if a great oscillation width is present, one of the populations might shrink and come under the threat of extinction due to a random event. If this population represents the predator, then the prey will have the opportunity to increase unobstructed, whereas if it represents the prey, the predator population will be eliminated due to the absence of food. However, in natural ecosystems, relations are multilateral, not bilateral; thus the prey population might be reduced by other factors or an alternative food source might be available to the predator.

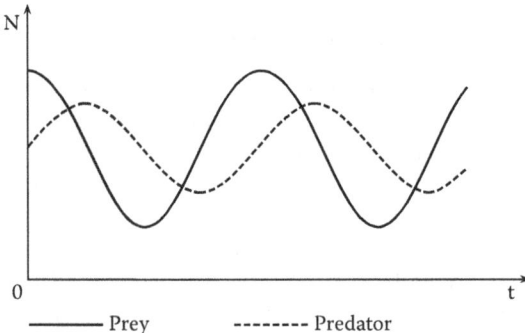

Figure 3.7 Predator-prey interaction. (From Hadjibiros 2007, modified. *Ecology. Ecosystems and Environmental Protection*, 3rd edition. Symmetria, Athens. With permission.)

Trophic relations, and therefore predator-prey interactions, are frequently successional in nature leading to the formation of the so-termed food chains. Usually, food chains combine with one another, generating complex trophic networks in which one population can be preyed upon by numerous predators and/or, simultaneously predate upon numerous preys. Primary producers and top predators are placed at the food chain boundaries.

3.4.3.1 Predator-Prey Interaction Models

Predation as a quantitative interaction of two populations is described by the typical Lotka-Volterra equations:

$$dN_1/dt = r_1N_1 - \alpha_{21}N_1N_2$$

$$dN_2/dt = -r_2N_2 + \alpha_{12}N_1N_2$$

which are based on the exponential model; they can be formulated for the logistic model respectively:

$$dN_1/dt = r_1N_1 - \alpha_{21}N_1N_2 - q_1N_1^2$$

$$dN_2/dt = -r_2N_2 + \alpha_{12}N_1N_2 - q_2N_2^2$$

where r_1 represents the prey increase rate when no predator is present and $-r_2$ is the predator decrease rate when no prey, i.e. no food, is present. The intensity of the populations' interaction is regulated by coefficients α_{21} and α_{12}. The presence of the N_1N_2 product is explained as in the case of competition.

From these equations it becomes obvious that prey growth rate decreases due to predator presence, whereas predator growth rate increases due to prey presence. We observe that if $N_2 = 0$ (predator absence), prey population N_1 increases. But if $N_1 = 0$ (prey absence), predator population N_2 decreases since the predator presumably has no other food source but the prey. In the case of an alternative food source existence, the second equation is rewritten without the presence of the (–) sign in front of the r_2N_2 term.

In the above equations the following assumptions have been made, which allow for a simple approach with enough realism, that is, without excessively departing from reality:

- Prey numbers consumed by the predator are proportional (linear relationship) to the prey population size.
- Predator population variation is proportional (linear relationship) to the number of preys consumed by it.

The substitution of the prey's linear response in relation to the predator $(-\alpha_{21}N_1N_2)$ and of the predator's linear response in relation to the prey

$(\alpha_{12}N_1N_2)$, with other more complex relationships might be useful when appropriate experimental data are present. Otherwise, it is preferable to remain with the linear relationship, which retains certain validity, at least as an initial approach. The above equations also presuppose that predator-prey interactions generate instantaneous consequences for both populations. This may apply more for mortality than for natality, the increase of which needs some time in order to materialise. Another possibility is the introduction of population age structure through models based on the use of matrices; however, this constitutes a structural change that results in more complicated outcomes.

The above equations cannot be solved analytically, but with the application of numerical methods. The graphical representation of the system demonstrates periodical oscillations. These oscillations have the same period for both populations but exhibit a phase difference, i.e. the predator's increase or decrease follows the respective prey's change. Oscillation amplitudes depend on the initial circumstances. These amplitudes remain constant if the exponential model is applied (Figure 3.7), whereas they decline if at least one of the populations exhibits self-limitation (Figure 3.8). The system balances either at a specific point or at a range of values of N_1 and N_2; in the first case the equilibrium is indifferent (neutral) whereas in the second case it is stable.

In the case of stable equilibrium, the prey's environmental carrying capacity must be large enough to support the predator population. In the opposite case, the system is modified due to predator elimination, whereas the prey population's size tends to balance at the value of the environmental carrying capacity. (See Figure 3.9.)

Elimination of one of the populations can occur in the case of indifferent equilibrium, when the initial values of population sizes are far enough from the equilibrium point, so that their oscillations have large amplitudes, i.e. are approaching the horizontal axis. This means that often, one of the

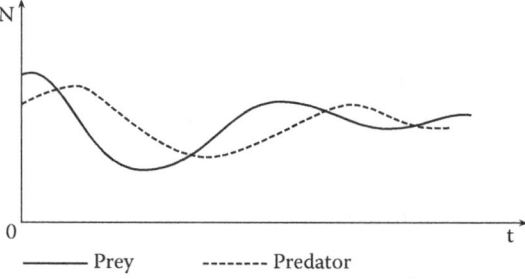

Figure 3.8 Predator-prey interaction with self-limitation. (From Hadjibiros 2007. *Ecology. Ecosystems and Environmental Protection,* 3rd edition. Symmetria, Athens. With permission.)

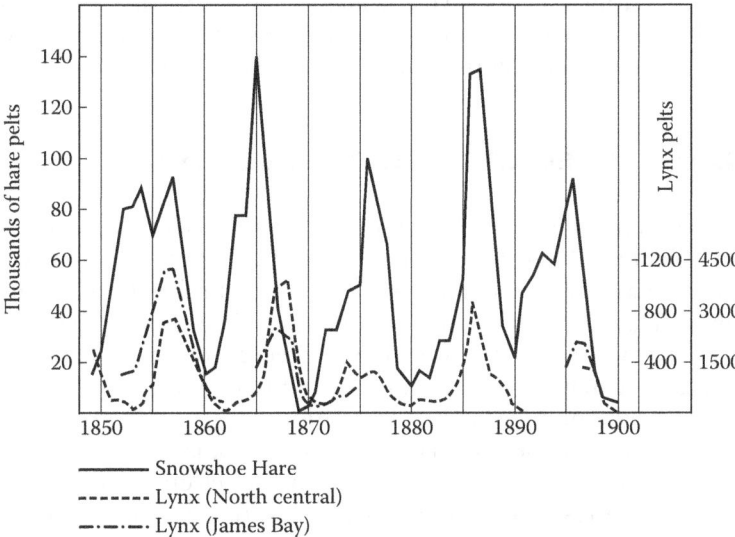

Snowshoe Hare

------- Lynx (North central)

—·—·—· Lynx (James Bay)

Figure 3.9 Relationship of hare and lynx populations in the Canadian arctic. Data are from sales records of traders dealing in both hare (*Lepus americanus*) and lynx (*Lynx canadensis*) skins. Lynx records are from two nearby trapping regions. The data show not only that the populations are linked but that numbers of the predator follow numbers of the prey. (From Finerty, J.P., 1980. *The Population Ecology of Cycles and Small Mammals*, Yale University Press, New Haven, CT and Colinvaux, P. (1993). *Ecology 2*. Wiley, New York. With kind permission from Yale University Press.)

populations may reach a very small size, and for some random occurrence, not recover and become eliminated. Then, if the predator is eliminated, the prey will survive, but if the prey is eradicated, the predator will also be eliminated. In both cases the system will be altered.

In such circumstances, the concept of refuge can be a stabilising factor, offering preys the ability to protect themselves against predation while giving predators the possibility of protection from exhaustion due to lack of food. Refuges can be spatial, for example a prey's shelter; temporal, as for example the existence of some biological stages in the prey's or predator's life span that do not contribute to predation.

The classic Lotka-Volterra model presents a tendency towards an oscillation with a phase difference, a common occurrence in natural predator-prey systems. The oscillation mechanism operates as follows: an increased prey population tends to provoke a growth in predator population, which leads to a decrease in prey population, leading to a decrease in predator population, which tends to reinstate an increased prey population and so on.

Simple predation models have a strategic character. An essential qualitative result is that predation causes population sizes to oscillate, however it does not generally cause extinction of one of the populations.

3.5 NATURAL SELECTION AND EVOLUTION

Evolution of a population is the total of successive alterations of the genetic information, i.e. of the population's hereditary characteristics' changes. A basic mechanism that causes evolution is natural selection that is the differential reproduction of the genetic types. The genetic characteristics of the population are not always reproduced according to their existing frequency; some may be favoured and reproduced more, thus increasing the frequency of their presence in the population.

An example of evolution observed in insects' populations is their increased resistance to insecticides. In this case, the natural selection mechanism functions under the pressure of mortality caused by the insecticide. The individuals of the insect population vary in relation to their resistance to the given insecticide, and this variance has a genetic, and therefore hereditary, basis. If the insecticide is applied, the most resistant insects have a greater probability for survival, and consequently for reproduction, that is, increased fitness for the given environment that includes the insecticide. Thus, since the hereditary quality of increased resistance will be transferred to more and more descendants, the population will become more and more resilient, with each successive generation, to the given conditions of the environment.

Another characteristic example is the continuous increase of the average volume of the human brain, which was observed in paleontological findings. In the human populations there was some variety in brain volume. Probably the individuals with the largest brains were on average more capable of using information, and this made them more capable of surviving, of finding a mate for reproduction and raising children. Therefore the number of fertile descendants, that is, their fitness for the environment, was increased in comparison to individuals with smaller brains. The result was the increase of the average brain size of human populations during prehistoric periods.

We conclude from the above that natural selection acts only if genetic variety pre-exists in the population. In populations without genetic variety, no differential reproduction of genetic types is possible. Such a population would find it hard to survive in the long run in an environment that would change, given that it would not have the ability to make evolutionary adaptations in the changing conditions.

A basic mechanism that creates new genetic variety in the population is mutation; that is, the accidental sudden alteration of the genetic information. Most mutations are harmful or even lethal, in which case they are rejected by the natural selection mechanism and become lost because they do not produce viable offspring. Rarely, though, some viable or even beneficial mutations occur which increase the species' fitness and can successfully lead to the creation of new species.

In each one of the "predator-prey" or "competitor-competitor" systems, there may be mutual pressure for evolution through the natural selection mechanism. The predator, for example, compels the prey population towards differential reproduction, which results in gradual improvement of the mechanisms that the prey possesses in order to avoid predation. Conversely, this improvement pushes the predator to new improvements of the attack mechanisms against the prey, again through natural selection. These evolutionary changes that influence each other constitute a co-evolution phenomenon.

3.6 SURVIVAL STRATEGIES OF POPULATIONS

Population is the ecological system that, in the frame of the science of ecology, corresponds to the lowest level of description. The organisms, the tissues or the cells do not constitute ecological systems because contrary to population, their elements are not autonomous but specialized, are not found in random positions, and they cannot distance themselves in space. Moreover, the organisms, the tissues or the cells have boundaries beyond which they cannot develop.

The temporal quantitative population alteration is effected under the influence of the environment that is manifested with the presence of food, natural conditions, disasters, predation, competition etc. At the same time, the population undergoes a slow qualitative change through the action of the environment in the form of natural selection on the population's individuals. The aim is the maximum increase of its size, with the result that the population forms a survival and growth strategy.

The study of a population's adaptation in a stable environment highlights the connection of some characteristics of the individuals with characteristics of the environment, such as:

- The relation of the individuals' size with temperature (a big size shows adaptation to the cold due to an improved relation of surface/size of the body)
- The relation of the individuals' colour with the substratum
- Predation behaviour in relation to the prey type

Other characteristics are adaptation to a changing environment that may, for example, change in time (seasonal periodicity) or exhibit spatial differentiations, fluctuations, uncertainty etc.

The population has to develop various strategies in order to cope with the demands of the environment. For example, an effort for adaptation to the alternations of a warm–cold environment may consist of an intermediate size of the individuals, in successful adaptation to one extreme only (e.g. to the cold environment), in polymorphism of the individuals with a different degree of adaptations for each one, in changing the size of the individuals according to the season etc.

Any survival strategy has a cost for the population, which is finally reduced to energy expenditure. For this reason, the developed mechanism should substantially increase the population's fitness and the ensuing benefit must be greater than the cost of the mechanism in order to be eventually advantageous.

In the end, the value of a survival strategy is a function of the relation "development to cost." This value will form a natural selection criterion of the specific strategy and will define the final success of the endeavour, i.e. the population's survival.

The strategies are connected, inter alia, with parameters as the population's natality and mortality. The available energy can be used by priority either for the increase of offspring production or for the enhancement of every organism's size and viability. In the first case, this energy is mainly invested in increase of natality, and in the second, in decrease of mortality. The population dynamics approaches, especially the exponential and the logistic model, set off two basic parameters:

- Parameter r, the intrinsic rate of natural increase. It is the current rate and not the maximum possible.
- Parameter K, the carrying capacity of the environment. It is not the highest limit, but an equilibrium point.

These fundamental population parameters refer to a specific period of time, are independent from the population's state and may depend on environmental conditions. They contribute to the determination of two basic survival strategies of populations:

- Strategy r: r's increase tendency, investment in natality
- Strategy K: K's increase tendency, investment in viability

Every ecological manifestation or behaviour can be considered as an r or K strategy element, as shown by these examples:

r: Generally small size.
K: Generally big size.
r: Energy investment in the population's biomass increase through reproduction. The population's survival is achieved even after a disaster (e.g. 99% deaths).
K: Energy investment in reserves and protection mechanisms of the individual against adverse environmental effects. The population survives only if an adequate number of individuals survive.

r: A great number of offspring. Short life span, many reproductive generations; increase of fertility span in relation to life span; easy propagation of seeds; opportunistic behaviour. Characteristic representatives: many insects, small mammals, weeds, many small seeds (e.g. fig tree) or eggs, etc.

K: Fat or nutrient reserves, hibernation, durable seeds, long life span, nursing, nuptial parades (reproduction of albatross or other birds), grooming of offspring; complex defence systems against enemy attacks; learning by experience; social behaviour. Characteristic representatives: big serpents, birds, mammals, species with shells or fur, big long-lived trees, acorns etc.

The speed of evolution is connected with the survival strategies:

r: Rapid evolution, favours the creation of new species

K: Slow or nonexistent evolution (perfected species, e.g. big trees, shark, albatross, bear)

The first settlers in a new environment (ecological void) are favoured if they have an *r* strategy and are opportunistic populations with a great migrating ability. They take advantage of the absence of hard competition and take possession of the ecological niche very quickly. They can deal more easily with the changing or unstable environment and unexpected phenomena. If needed, they migrate elsewhere. Towards the end of ecological succession, the permanent, mature inhabitants will settle in the environment. They take advantage of the prevailing stability and the predictability of changes and they have learned to be effective in the exploitation of resources. They can deal with intense competition and they gradually prevail. They cannot recover easily if the population is greatly reduced, and that's why they need more protective care.

In nature, *r* and *K* strategies constitute the extreme cases of a continuous range of intermediate strategies. Prevalence of the first or the second in a greater degree contributes to the correspondingly lesser or greater stability of the ecosystems.

In human populations, *r* and *K* strategy elements occur such as polygamy—the subordination of women and their utilization in the maternal role—on one hand, and the long-term raising of children, the emancipation of women, and the development of civilizations on the other.

3.7 THE HUMAN POPULATION

The population of the human species constitutes a special case in the biosphere's history. Man significantly modifies the natural environment;

he has been an accomplished predator of many—and still is a redoubtable competitor of many more—biological species. Human overpopulation constitutes the main factor of degradation of the biosphere and the environment in general (Chapter 5). As early as 1798, economist Thomas Malthus suggested that the increasing population will cause depletion of the resources, which will lead to poverty, hunger and death. Although the Malthusian predictions have not been confirmed because of technological and social activities (Chapter 13), the great increase of the global population during the last decades, the generalized expectation of approaching a high level of consumption and the use of harmful productive methods (Chapters 7, 8, and 9) lead to an increase of pressures on the environment and the resources. An observation of human population history reveals the factors that influence its evolution.

3.7.1 The Changes of Human Populations

Historical data about the sizes of human populations exist for the period after AD 1650. Population estimates about previous periods are indirect, with few circumstantial facts. Before 8000 BC, for example, agriculture had not developed, man was a wild animals' hunter and fruit gatherer and the estimate is based on the observed density of hunters-gatherers' groups today. For the period from the development of agriculture until the middle of the 17th century, when population census begun, the estimates were performed based on observations of contemporary rural societies in conjunction with archaeological findings (Table 3.1).

Table 3.1 Demographic History of Humanity

Date	Cultural Evolution	Global Population (in millions)
−1000000	Lower Palaeolithic era	0.125
−100000	Middle Palaeolithic era	1
−25000	Upper Palaeolithic era	3.5
−4000	Neolithic era	26.5
−100	Classic Antiquity	133
1650	Early Modern Times	545
1800	Beginning of Industrial Era	900
1950	Atomic Age	2400
1960	Space Exploration	3000
1970	Computer Science Development	3660
2000	Globalization	6071
2030 (prob.)	Post-Industrial era	8100

Sources: Modified from Hadjibiros, K. (2007). *Ecology. Ecosystems and Environmental Protection,* 3rd edition. Symmetria, Athens (in Greek); and Ramade, F. (1974). *Elements d'Ecologie Appliquée.* McGraw-Hill, New York. With permission.

Table 3.2 Global Population and Increase Percentages

Date	Population (millions)	Mean Yearly Increase (percentage %)	Doubling Time
–8000	5		
		0.0477	1453
1650	500		
		0.3466	200
1850	1000		
		0.916	76
1950	2500		
		1.880	37
1975	4000		
		1.887	37
1988	5112		
		1.433	49
2000	6071		
		0.787	88
2050 (prob.)	9000		

Source: Hadjibiros, K. (2007). *Ecology. Ecosystems and Environmental Protection*, 3rd edition. Symmetria, Athens (in Greek). With permission.

Table 3.2 juxtaposes global population in some characteristic dates with the average yearly increase rate in the intermediate periods, based on an exponential model. It also shows the theoretical times of population doubling for each intermediate period, which constitute a characteristic index of the increase.

According to archaeological findings, agricultural communities had begun to function in several sites around 10,000 BC. With the beginning of the agricultural period, the increase of human population was accelerated, mainly due to a decrease of mortality. The development of agriculture gave man enough food and allowed him to settle in one place. It also gave to members of the rural society the possibility to turn to other activities that assisted the rise of the general life standard. Thus, the natural dangers and other negative factors for human life were eased off; as a result, the average life expectancy rose above the initial level of 20–25 years. According to other views, the increased acceleration is rather due to the increase of natality than to the decrease of mortality. Man, as a moving hunter and fruit gatherer, left perhaps long time intervals between births, which were imposed by the mother's inability to carry along more than one child as well as by the need for nursing for at least 3 years, since the available hard food was unsuitable for very small children. Therefore, one of the most "human" characteristics, birth control, probably appeared many thousands of years ago. On the contrary, during the agricultural period, children were not a burden any more but constituted a capital with economic value for the cultivation of the land.

During the agricultural period, until the middle of the 17th century, a slowly accelerating population increase is observed, although in the short run there are regressions. A characteristic such case is the decrease of the European population from 85 to approximately 60 million in the middle of the 14th century due to bubonic plague (black death). From 1348 to 1379, the English population fell from 3.8 to 2.1 million due to this plague.

The pre-industrial period, 1650–1850, is characterized by a great acceleration of the human population increase. It is estimated that the yearly percentage of increase was 0.3% in the century 1650–1750 and 0.5% in the century 1750–1850. During the first century, the population of Europe and Russia increased from 103 to 144 million, while during the second the population of Europe doubled. In that last period the population of the New World increased from 12 to 60 million, something that was to a great part due to immigration. The population of the United States was, according to the first census (1790), 4 million, while in 1850 it reached 23 million.

The rapid increase of agricultural production and of trade, the New World's opening to European exploitation and the adoption of personal hygiene habits were causes that influenced directly or indirectly death and birth rates. The introduction of two new foodstuffs, maize and potato, definitely influenced the population increase, given that with these foodstuffs a large family could be supported from a small piece of land. The wide application of potato cultivation in Northern Europe during the 18th century was probably the main reason for the population increase in that area. Ireland's population explosion is characteristic; during that period, the population increased from 3 to 8 million. Also, a great population decrease in Ireland followed the destruction of potato production in 1846 and 1847 due to a plant disease. It is estimated that one million people died, while two million people immigrated. In Asia, the population increased by 50–70% during the period 1650–1750. It seems that this increase took place mainly in China as a result of a progressive agricultural policy and the introduction of maize and other new vegetable foods. In the period 1750–1850, the population of Asia increased by approximately 50%. Africa's population during the period 1650–1850 seems to have remained almost stable, around 100 million.

The industrial revolution, in spite of the terrible, by the current criteria, working conditions in mines and factories, caused an immediate rise of the standard of living in Europe and North America. The progress in agriculture, industry and transportation, the sanitary technology works (sewers, drinking water purification) during the last third of the 19th century, and the introduction of vaccinations constituted mortality decrease determinants. In Europe, the percentage of mortality was 22–24‰ in 1850 and

18–20‰ in 1900, while in Sweden, Denmark and Norway it fell from 20‰ in 1850 to 16‰ in 1900. In Western and Northern Europe, the population increase led to mass immigration.

Around the end of the 19th century, a new significant trend appeared in some European countries and in the United States, namely the diminution of natality. It was the beginning of the so-called demographic transition, that is, of a diminution of natality that follows the diminution of mortality, and which constitutes a phenomenon that accompanies industrialization, although significant deviations are observed. Generally, demographic transition is associated with a diminution of natality because as a society becomes industrialized and agriculture is automatized, a child ceases to be an economic gift for the family, especially after the prohibition of work for minors and the institutionalization of compulsory education. Moreover, a large family does not favour the accumulation of capital. Significant factors are the woman's education and work as well as the weakening of the traditional religious influence. In some countries, e.g. Hungary, the diminution of natality began quite before industrialization. France was the first country where diminution of natality was observed (circa 1800), while its industrialization was delayed in comparison to its neighbours. Demographic transition continued in the developed countries during the 20th century. Nevertheless, their population significantly increased during the period 1850–1950. After World War II, a natality increase (baby boom) was initially observed in the developed countries, and subsequently a decrease that reached a zero or negative population change. In many countries, the Welfare State took measures in order to boost natality.

In non-developed countries, especially after World War II, a rapid change in the dynamics of human population—the post-war population explosion—took place. With the introduction of medicines and public health measures a dramatic decrease of mortality was realized. For example, life expectancy at the moment of birth increased from 41 years in 1950 to 63 years in 2000 in the less developed countries. This decrease, however, was not accompanied by an immediate natality decrease, given that socioeconomic and educational conditions did not change significantly during the same period in most of the non-developed countries. It is characteristically reported that the birth rate in India in the years 1891 and 1931 was 49/1000 and 46/1000 respectively.

In the 20th century, a great improvement in the sector of human health was achieved, with the result that the population of developed countries more than doubled and that of the developing countries quadrupled and came close to five billion. At the same time, a natality decrease is observed, which spreads to many developing countries and is mainly due to the spreading of women's education and work (Figure 3.12), as well as to the success of family planning programmes. Such programmes began to develop, for example, in India, since 1952, but it took approximately three decades in

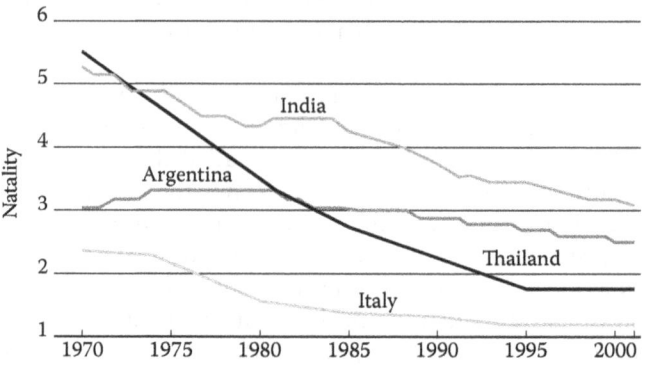

Figure 3.10 Natality decrease in developed and developing countries. (From Hadjibiros 2007, *Ecology. Ecosystems and Environmental Protection*, 3rd edition. Symmetria, Athens www.unfpa.org. With permission.)

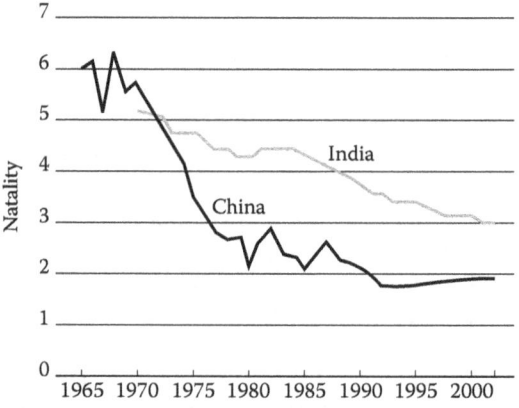

Figure 3.11 Natality decrease in China and India. (From Hadjibiros 2007, *Ecology. Ecosystems and Environmental Protection*, 3rd edition. Symmetria, Athens www.unfpa.org. With permission.)

order for the demographic policy to become generalized. Natality in less developed countries began to significantly decrease during the 1970s and the 1980s (Figures 3.10 and 3.11). The result was a natality decrease below the minimum limit for population reproduction (2.1 children per woman) in almost all developed countries, but also in China, Brazil, Korea, Thailand and other countries that correspond to one fourth of the developing countries' total population. The countries that maintain very high birth rates do not represent more than 5% of the actual global population.

A factor that significantly influences the population of many countries is international migration. It is estimated that more than 200 million people

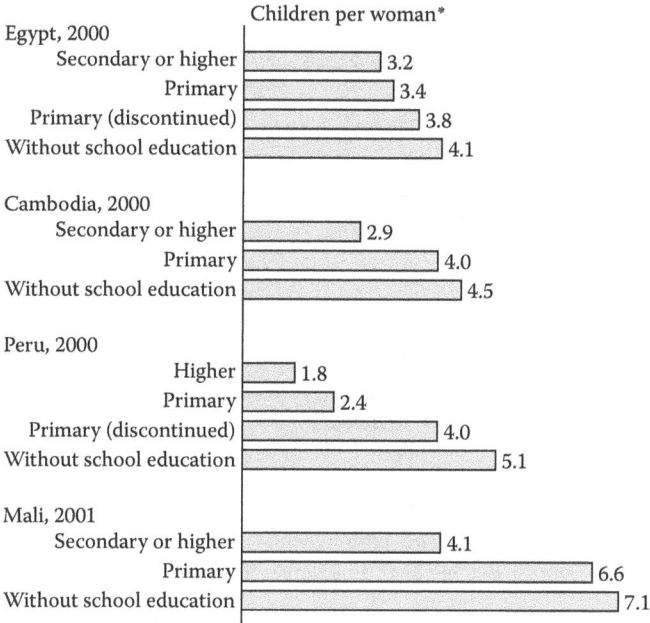

* Mean number of children on the basis of natality.

Figure 3.12 Relation between natality and educational level of mothers. (From Hadjibiros 2007, *Ecology. Ecosystems and Environmental Protection*, 3rd edition. Symmetria, Athens www.unfpa.org. With permission.)

were foreign immigrants in 2006 and their number continues to increase. They usually come from developing countries and less from developed ones and they immigrate to developed or to some developing country where employment opportunities are better. One factor that plays a significant role in the mortality rate of some developing countries is the diseases that are not combated effectively, such as AIDS. In Europe, the age structure of the population has been considerably influenced by factors such as the high mortality and low natality rates during World War II, the "baby boom" after the war, the long period of low natality that followed and the inflow of immigrants during the last decades. It is estimated that in order for the European population reduction to be avoided, a yearly immigrants' inflow twice that of 1990 would be needed.

The most probable course is that the global population will be stabilized after a few decades (Figure 3.13). It is considered very probable that natality will continue to decrease in those developing countries where such a tendency exists today and it is expected that the decrease will also take place in countries where natality has remained consistently high. Urbanization and the modernization of societies in developing countries are considered

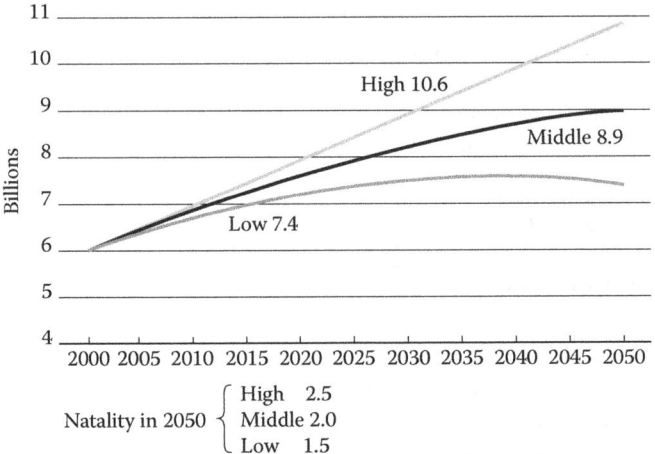

Figure 3.13 Predicted human population sizes: three scenarios. (From Hadjibiros 2007, *Ecology. Ecosystems and Environmental Protection,* 3rd edition. Symmetria, Athens www.unfpa.org. With permission.)

to be non-reversible tendencies. A central question, however, is if and to what degree the poorer and uneducated, mainly rural, groups will adopt the small family model that urban societies follow, a model that presupposes a higher educational and economic level. For the time being, a large and increasing number of people lack the necessary education, health and family planning services.

Chapter 4

Organisation at the Ecosystem Level

DOI: 10.1201/b14609-4

4.1 ENERGY FLOW IN THE ECOSYSTEM

Energy flow represents a fundamental ecological concept with a unifying role among the partial plant, animal, and microorganism studies. It is based on trophic relations between various organisms and constitutes the "driving force" for the function of the ecosystem. Energy flow is not circular but linear (Figure 4.1), with a direction from the lower to the higher trophic levels and with conversion from light to chemical energy and, finally, from chemical energy to heat, i.e. to a downgraded energy form. Energy flow interruption in an ecosystem entails the collapse of its organization; the ecosystem is gradually decomposed into inorganic matter.

The energy source for the biosphere is solar radiation. Only a few special deep sea ecosystems represent exceptions, since these can obtain energy from chemical substances without the presence of light. The manifold life manifestations depend on light for energy. Plants and algae and some bacteria bind and exploit light energy through photosynthesis.

Solar radiation with intensity $0-1.2$ cal/min/cm^2 reaches the planet Earth. The main part of this energy mobilizes wind, wave and oceanic current mechanisms as well as the hydrological cycle, ultimately returning to space as infrared radiation. It is estimated that around 1% of the total incident radiation is bound by the biosphere through photosynthesis.

Two essential concepts used for the study of energy flow in an ecosystem are biomass and production. Biomass is defined as the energy content (or the equivalent organic matter content) of an organism, population or ecosystem. Production is the amount of biomass increase, while productivity is production per time period. Biomass and production are measured, as a rule, in energy units (e.g. calories) or mass units (e.g. grams). Productivity is usually expressed in mass units per time and surface.

Other useful concepts are consumption, assimilation, waste and respiration, which are measured by the same units. Hence, trophic relationships are analysed as follows: the biomass deducted from a prey population, i.e. from an inferior trophic level, serves as consumption to a predator population,

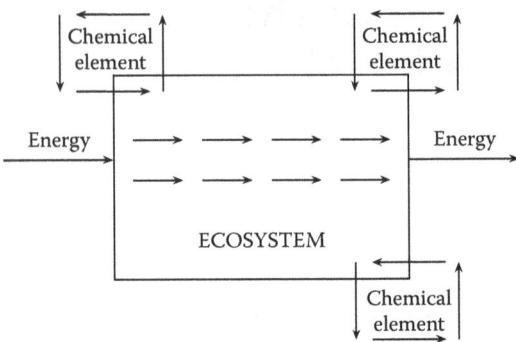

Figure 4.1 Energy and matter flow. (From Hadjibiros, K. (2007). *Ecology. Ecosystems and Environmental Protection,* 3rd edition. Symmetria, Athens (in Greek). With permission.)

i.e. a superior trophic level. A part of this biomass is utilized and the rest is discarded; consequently, the consumption amount can be analysed in assimilation and waste. From the assimilation amount a part is ultimately transformed in heat and discarded through respiration, whereas the rest constitutes the trophic level production. In its turn, this can be utilized as food for a subsequent trophic level.

According to the above, the energy magnitudes of a population or ecosystem are connected through the following relationships:

$$C = A + W$$

$$A = R + P$$

where C is consumption, A assimilation, W waste, R respiration and P production.

4.1.1 Food Chains—Trophic Levels

On the basis of their position in the ecosystem from an energy flow aspect, organisms are divided as follows:

- Producers are the autotrophic-photosynthetic organisms. These have the fundamental capability to produce synthetic organic matter using inorganic matter along with solar energy. Plants are the producers in terrestrial ecosystems, while in aquatic ecosystems they are mostly algae.
- Consumers are heterotrophic-chemosynthetic organisms, apart from bacteria and fungi. They are only capable of composing organic matter after obtaining carbon and energy from the producers.

- Decomposers are the bacteria and fungi which are consumers in essence but execute a special function in the ecosystems. Their role is to break down dead organic matter to inorganic compounds, which thus become available again for producers' consumption.

A series of species, each one of which constitutes food for the next (Chapter 3, Section 3.4.3), is described as a food chain. Often, food chains are interconnected, since many ecosystem organisms belong to more than one food chain. Thus, trophic webs, which can be divided in trophic levels, are created. The first trophic level contains plants or algae, the second one herbivores or zooplankton, the third one the first carnivore animals, the fourth the second carnivore animals etc. Omnivorous animals are placed in more than one trophic level.

During energy flow from one trophic level to another, the available energy is reduced (Figure 4.2). In each step, a small part of the incoming energy equivalent of food is discarded as waste, while the main part is used for the energy needs of the respective level; the bigger part of it is discarded as heat through respiration, while the rest is integrated as chemical energy to the produced biomass. Energy efficiency (according to Lindeman) is the

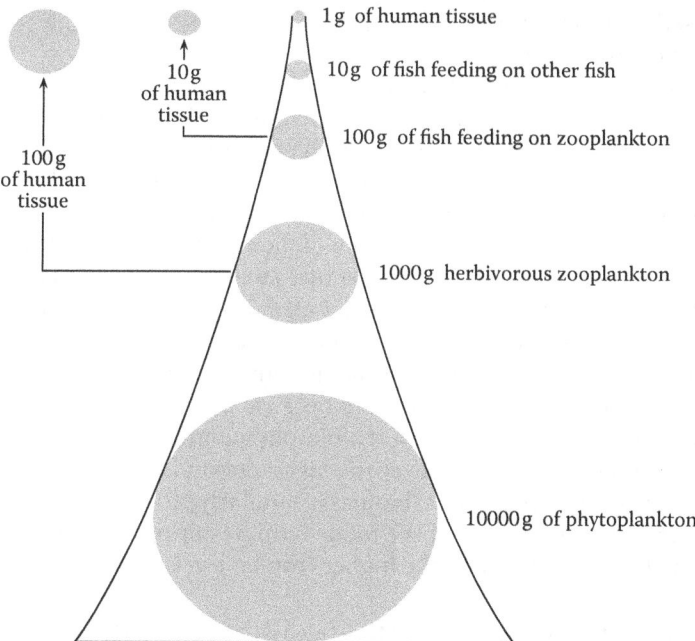

1g of human tissue

10g of human tissue

100g of human tissue

10g of fish feeding on other fish

100g of fish feeding on zooplankton

1000g herbivorous zooplankton

10000g of phytoplankton

Figure 4.2 Pyramid of masses. (From Hadjibiros, K. (2007). *Ecology. Ecosystems and Environmental Protection,* 3rd edition. Symmetria, Athens (in Greek). With permission.)

values' ratio of the chemical energy assimilated in two successive levels n and $n - 1$:

$$a = \frac{A_n}{A_{n-1}}$$

or

$$a = \frac{R_n + P_n}{C_{n-1} - W_{n-1}}$$

Generally, in the biosphere, efficiency ranges on average around 10%, even though much greater or smaller values can be observed. Thus only about 10% of the chemical energy entering a trophic level is available to the next level. Consequently, in each successive step, 90% of the energy is lost in the form of heat or waste. It is a fact that high energy efficiency is not a characteristic of natural ecological phenomena.

The great energy losses do not allow the creation of long food chains in nature. The number of trophic levels usually ranges between 3 and 5; it can reach up to 8 in some special cases. Usually, the final link of a complete food chain, i.e. the highest predator, has a small population size and suffers an increased elimination risk. Subsequently, the presence of highest predators is an indication of a healthy ecosystem.

Great energy losses imply that the available food in an ecosystem for an omnivorous consumer, such as man, is much more affluent in lower trophic levels than in higher ones; thus, from an energy ecological point of view, it is much more expedient for a human population to consume cereal than beef.

The phenomenon of energy loss along the food chain is responsible for an important ecological mechanism known as biomagnification that increases the toxic consequences of polluting substances which cannot be easily broken down (Chapter 7). When, for example, a chemical substance which cannot be broken down appears in the marine phytoplankton with concentration a, its concentration in the zooplankton (phytoplankton's predator) will be $10\,a$ if energy efficiency is 10%. This occurs because in order for a zooplankton biomass quantity to be produced, a tenfold phytoplankton quantity must be consumed and therefore the stable chemical compounds will be accumulated proportionally in the predator's biomass. Similarly, this process will take place in each of the next steps of the food chain, resulting in a great accumulation of the toxic substance in the higher trophic levels.

4.1.2 The Role of Photosynthesis and Respiration

Photosynthesis is the basic production process supporting Earth's ecosystems from an energy point of view. Only in some special sea-bed areas near geothermic sources can certain bacteria support trophic chains by producing organic matter from inorganic compounds, without light. Thus, almost all biosphere production, but also all atmospheric oxygen, in addition to mineral fuels, are the products of photosynthetic processes. Primary production is the created biomass of producer organisms, whereas the production of consumer organisms is called *secondary production.*

Respiration is the basic ecological process of energy consumption. It corresponds to the total of chemical reactions of all the aerobic organisms' metabolism taking place in the ecosystem. Photosynthesis and respiration are the biological phenomena determining energy flow, as well as the circulation of carbon and oxygen, in the biosphere.

4.1.3 Ecosystem Energy Model

A significant way of studying and understanding ecosystems' function is the analysis of matter and energy exchanges between ecosystems and their environment, as well as the analysis of energy and matter flow between different sectors of the ecosystem. Useful mathematical tools for this analysis are multi-compartment models. The basic idea consists in the division of the system in a finite number of compartments, interconnected with quantitative exchanges. The system state is determined by the content values of all the compartments. The system's time variation depends on flow values between the compartments. These models have been widely used for the study of ecosystems' energy state.

The general picture of energy flow in any ecosystem or in the whole biosphere is given by a simple model (Figure 4.3), with variables:

B_1: producers (plants or algae)
B_2: primary consumers (herbivorous animals, zooplankton)
B_3: secondary consumers (carnivorous animals)
B_4: decomposers
L: absorbed radiation through photosynthesis
R_μ: respiration of compartment μ
$F_{\nu\mu}$: flow between compartments ν and μ

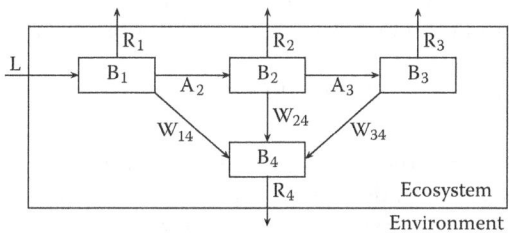

Figure 4.3 Energy model of an ecosystem. (Modified from Hadjibiros, K. (2007). *Ecology. Ecosystems and Environmental Protection,* 3rd edition. Symmetria, Athens (in Greek). With permission.)

F_{12} and F_{23} are A_2 and A_3 (assimilations of compartments 2 and 3), respectively and F_{14}, F_{24}, and F_{34} are W_{14}, W_{24}, W_{34} (wastes from compartments 1, 2, and 3 to compartment 4), respectively.

The ecosystem's state is determined by the energy content of the four compartments, which can correspondingly be expressed as biomasses: B_1, B_2, B_3, B_4.

System dynamics are described by the following equations:

$$dB_1/dt = H - R_1 - F_{12} - F_{14}$$

$$dB_2/dt = F_{12} - R_2 - F_{23} - F_{24}$$

$$dB_3/dt = F_{23} - R_3 - F_{34}$$

$$dB_4/dt = F_{14} + F_{24} + F_{34} - R_4$$

A simulation of the system's behaviour requires some assumptions on various variables and relations. For example, it may be assumed that incoming solar radiation is $L = \alpha$ (fixed variable) or $L = \alpha B_1$ or $L = \alpha \eta \mu (\omega t)$ etc. Flows can, in an initial approach, be considered analogous to the contents of the sections they originate from: $F_{\mu\nu} = g B_\mu$ (linear hypothesis), as, e.g., in the case of flow towards the decomposers, that may be proportional to mortality. If flows are considered analogous to the contents of the two compartments they connect, then these flows correspond to predation interactions, e.g. between B_1 and B_2 or B_2 and B_3, in which case the relation $F_{\mu\nu} = g B_\mu B_\nu$ corresponds to Lotka-Volterra equations (Chapter 3). Generally, flows can represent various interactions between the corresponding populations or be dependent on other variables as well.

The results of system simulation can be assessed on the basis of general criteria, such as: the existence of a stable equilibrium point which is relatively soon reached by the system, the limitation of the system's sensitivity in parameter's change etc. Given that this kind of strategic model cannot give accurate representations of reality, it is preferable for the assumptions on which they are based to be relatively simple.

Biomass variation or energy variation of the ecosystem is:

$$E = dB_1/dt + dB_2/dt + dB_3/dt + dB_4/dt$$

If $B_1 + B_2 + B_3 + B_4 = B$ and $R_1 + R_2 + R_3 + R_4 = R$ and given that, on the basis of the above ecosystem model, energy balance is: $E = L - R_1 - R_2 - R_3 - R_4$, the variation of the ecosystem biomass is equal to the difference of photosynthesis from the total respiration:

$$E = dB/dt = L - R$$

Thus, energy flow from the ecosystem to the environment and reverse is mainly dependent on the basic chemical reactions of photosynthesis and respiration:

$CO_2 + H_2O +$ radiation $\rightarrow$ organic matter $+ O_2$ (photosynthesis)

Organic matter $+ O_2 \rightarrow CO_2 + H_2O +$ heat(respiration).

It is obvious that the relationship between incoming energy (radiation) and outgoing energy (heat) results in a corresponding relationship between the incoming and outgoing CO_2 quantity as well as between the incoming and outgoing O_2 quantity. Thus:

- If $E = 0$, then the ecosystem is in energy balance with its environment and its total biomass does not change. The CO_2 quantities absorbed and emitted by the ecosystem are equivalent, and the same is true for the absorbed and emitted O_2 quantities. This represents the case of an ecosystem in a condition of stable equilibrium, as e.g. a mature forest.
- If $E > 0$, the ecosystem is an O_2 producer and CO_2 consumer. The absorbed CO_2 quantities are greater than the emitted ones, whereas the opposite is true for O_2. In this case there is an energy inflow in the ecosystem and its total biomass increases; therefore it is in a developmental phase.
- If $E < 0$, the ecosystem is an O_2 consumer and CO_2 producer. The absorbed CO_2 quantities are less than the emitted ones, while the opposite is true for O_2. Energy balance is negative and its total biomass decreases. This is an ecosystem in a shrinking phase.

If $E > 0$ or $E < 0$ or $E = 0$, it does not necessarily follow that all biomasses of individual compartments are increased, decreased or remain stable, respectively.

From the above, it becomes evident that a mature ecosystem, i.e. an ecosystem that does not develop anymore because it has reached an energy balance, does not produce O_2 nor consumes CO_2. On the contrary, these results are mainly produced by developing ecosystems (e.g. a growing forest or a new forest settling in an open space) or ecosystems, part of the production of which does not decompose but is removed by man (e.g. agrosystems, forests under exploitation).

During some geological eras (e.g. carboniferous), Earth's ecosystem, i.e. the biosphere, was in $E > 0$ states. During such periods fossil fuel deposits were created, because the increasing biomass could not be decomposed and was stored in the ground; thus, the additional energy of the ecosystem was stored in the form of chemical energy of the fuel. On the other hand, there were situations when $E < 0$, such as the dinosaurs' extinction period or periods of extensive forest fires or of other analogous destructions.

The approach of the energy model can be expanded to include non-natural ecosystems. An interesting example is that of the urban ecosystem, where the function of machines and the rest of combustions are added to human and other organisms' respiration, increasing by a large amount the total O_2 consumption and CO_2 production within the cities. The urban ecosystem is continuously consuming O_2. The deficit is covered by air inflow from areas outside the cities, since O_2 production from urban area plants is negligible. Besides, biomass production within urban ecosystems is much smaller than their consumption needs, whereas their decomposers cannot consume the large amount of the created waste within these areas. Thus, the food deficit and the surplus of waste require imports of food and exports of waste, resulting in an urban ecosystem much dependent on the operation of other ecosystems located outside the cities (Figure 6.1).

4.2 LIMITING FACTORS IN VARIOUS ENVIRONMENTS

An important factor affecting biomass increase may, under specific conditions, become the critical factor that inhibits the production of one or more populations or trophic levels. Its limiting action may be connected with energy, nutrients, abiotic conditions or the influence of the biocoenosis. The idea that organisms' increase is controlled by the weakest link of the chain of their needs comes from Justus von Liebig (1840), a pioneer in the study of inorganic fertilizers and is known as the "law of the minimum." According to von Liebig, the amount of living matter in a given environment is limited by the factor which is characterised by the smallest, proportionally, availability, on the condition that no other negative factors interfere.

The limiting factor in the desert is H_2O, whereas in other terrestrial ecosystems it may be either H_2O or CO_2. In most marine ecosystems, where H_2O and CO_2 are available, primary production is usually limited by the elements P or N, which are needed in the form of salts. Some consumer organisms of the marine environment are limited by O_2. In the open ocean, the limiting nutrient may be iron (Fe).

A generalisation of the above approach can be expressed as follows: The development of a biocommunity (or a population) is dependent on a total of conditions and the one which comes closer (either upwards or downwards) to the corresponding tolerance limit of the biocommunity tends to become the limiting factor. This is a concept that can be very useful for the study of any ecosystem or part of it. The ultimate factor limiting biosphere's development is the incident solar radiation.

The limiting factor is defined in relation to some time scale. Generally, there is only one in each section and phase of the ecosystem. Yet if its availability increases, some other factor may become limiting. When conditions change, successional limiting factors appear, as for instance in the case when the values of factors fluctuate naturally. The existence of a limiting factor leads the respective population or part of the ecosystem to an increase with self-limitation (Chapter 3). The limit of this increase, i.e. the carrying capacity of the environment, is each time determined by the limiting factor. The possible absence of a limiting factor in a sector or phase of the ecosystem leads to an exponential increase state.

The notion of the limiting factor allows the comprehension of essential ecological phenomena, without the necessity of a detailed analysis of the structure and functions of the respective ecosystems. The value of this approach is due to the fact that the researcher can explore the ecosystem, thus "starting a thread" through complicated situations. However, this is not always applicable in transitional or swiftly changing ecosystem states, but mostly in steady states close to the equilibrium point.

4.2.1 Primary Production

Primary production is defined by the efficiency of producer organisms and essentially determines the productivity of the whole ecosystem (Table 4.1). It depends on the following factors, one of which becomes, in each case, the limiting factor:

- Radiation is the most important natural factor. Its daily and annual periodicity regulates the activities of a large number of organisms. It can become the limiting factor if its intensity is too low. Visible light intensity is a function of the geographical amplitude, climate, relief, orientation, water depth, transparency etc. In aquatic ecosystems, algae and aquatic plants do not develop in the absence of light, because photosynthesis is not possible and productivity is nullified. Therefore, light becomes the limiting factor in the dark depths of the lakes or the sea, where it cannot penetrate. The same happens to all ecosystems at night. During the winter, light may become a limiting factor if its intensity is reduced. Intense ultraviolet radiation can also become a limiting factor in terrestrial ecosystems (Chapter 9), since it harms the organisms.
- Temperature is also an important natural factor and can become the limiting factor when it reaches extreme values, e.g. in polar areas or during the winter. Temperature variations are generally greater in terrestrial than in aquatic ecosystems.
- Humidity, i.e. water, is a very important and often a limiting factor in terrestrial ecosystems (e.g. desert) whereas it cannot, obviously, limit aquatic ecosystems. In ecosystems such as wetlands water can become a regulative factor. Salinity may have a corresponding part in coastal wetlands. In terrestrial ecosystems, where water is not abundant, fire can become an important regulating factor.

Table 4.1 Primary Production and Biomass of Various Ecosystem Types in the Biosphere

Ecosystem Type	Surface (10⁶ sq km)	Primary Productivity (gr/m²/year)	Biosphere Primary Production (10⁹t/year)	Global Biomass (10⁹t)
Tropical rainforests	17.0	1000–3500	37.4	765
Deciduous tropical forests	7.5	1000–2500	12.0	260
Temperate forests	12.0	600–2500	15.1	385
Northern forests (Taiga)	12.0	400–2000	9.6	240
Scrubs and savannahs	23.5	200–2000	19.5	110
Tundra	8.0	10–400	1.1	5
Deserts and half-deserts with scrubs	18.0	10–250	1.6	13
Barren deserts and polar zones	24.0	0–10	0.07	0.5
Agrosystems	14.0	100–3500	9.1	14
Wetlands	2.0	800–3500	4.1	30
Lakes and rivers	2.0	100–1500	0.5	0.05
Total of terrestrial ecosystems			110	
Open ocean	332	2–400	41.5	1.0
Upwellings	0.4	400–1000	0.2	0.008
Shelfs	26.6	200–600	9.6	0.27
Coral reefs	0.6	500–4000	1.6	1.2
Estuaries	1.4	200–3500	2.1	1.4
Total of marine ecosystems			55.0	
Biosphere total			165	

Sources: Modified from Lieth, H. and R.H. Whittaker (1975). *Primary Productivity of the Biosphere.* Springer-Verlag, Heidelberg; and Ramade, F. (1974). *Elements d'Ecologie Appliquée.* McGraw-Hill, New York. With permission.

- Soil quality and the presence of nutrient salts and of trace elements are important. They are also significant for the proper fertilization of agricultural cultivations.
- The concentration of nutrient salts or trace elements in water constitutes the limiting factor that usually determines aquatic ecosystem's productivity.
- Carbon dioxide concentration in the air or the water is possibly the main limiting factor for the biosphere in recent geological periods and in a large time scale.
- Adverse effects of parasites, pollution, grazing or other anthropogenic interventions may be limiting factors.

An amount of anthropogenic limiting factors is associated with pollution. Here, one can discern the following:

- Substances that harm due to their excessive quantity, even though they are not foreign to natural ecosystems (e.g. dead organic matter in the water). If pollution does not exceed certain limits, ecosystems have significant assimilation ability, thus the presence of pollutants will not act as a limiting factor.
- Substances that harm, due to their toxicity, even in small concentrations (e.g. some heavy metals or radioactive substances). In such cases, pollution avoidance will abolish the limiting factor.

Generally, the most productive terrestrial ecosystems can be found in the Equatorial regions, where increased sunlight and humidity are present. On the contrary, the most productive marine ecosystems are often far from the Equator, in points where sea currents convene and, in particular, when there are upwellings (upward water movements) that bring nutrients from the bottom to the surface. Many coastal ecosystems are productive for the same reason.

Comparative examination of our planet's natural ecosystem's primary production illustrates the superiority of tropical rainforests, coral reefs, and some wetlands. Deciduous tropical forests, temperate coniferous or deciduous forests, cold northern forests and savannahs rank second; shrub, lakes, rivers, marine upwelling regions and coastal zones rank third; tundra, deserts and the deep ocean rank last with some deserts and polar areas at the extreme limit of low productivity. Agriculture is mostly placed approximately in the middle of this scale. Extremely productive cultivations, such as that of sugarcane, can be compared with the most productive natural ecosystems, although this only occurs due to great energy, fertilizers, pesticides, and water inflow from the outside.

Total solar energy converted into chemical energy through photosynthesis is the gross primary production for the whole biosphere. Deducting respiration from this amount, net primary production results; about two-thirds of this amount come from terrestrial ecosystems, whereas one-third come from marine ecosystems. Net primary production represents available energy for the biosphere's food chains.

4.3 BIOGEOCHEMICAL CYCLES

In each ecosystem an uninterrupted circulation of material substances is observed, in alternating inorganic and organic forms (Figure 4.1). The existence of an organism's groups with complementary needs within the ecosystem allows the recycling of the chemical elements. Various populations constantly seek and utilize the substances necessary for their increase, conservation, and reproduction and discard inorganic and organic waste in the

Table 4.2 Presence of Chemical Elements in Maize and Human Body (Dry Weight) and Availability of These Elements in the Abiotic Environment

Chemical Element	Presence Percentage in the Earth (lithosphere, hydrosphere, atmosphere)	Presence Percentage in Maize (dry weight)	Presence Percentage in the Human Body (dry weight)
Oxygen	46.68	44.57	37.13
Silicon	27.60	1.17	0.00
Aluminium	8.05	0.11	0.00
Iron	5.03	0.88	0.01
Calcium	3.63	0.20	3.75
Sodium	2.72	0.10	0.40
Potassium	2.56	0.92	0.90
Magnesium	2.07	0.18	0.10
Phosphorous	0.152	0.18	2.50
Carbon	0.149	43.70	45.00
Hydrogen	0.145	6.26	8.40
Manganese	0.116	0.00	0.00
Sulphur	0.100	0.17	0.60
Chlorine	0.095	0.14	0.40

Source: Modified from Hadjibiros, K. (2007). Ecology. Ecosystems and Environmental Protection, 3rd edition. Symmetria, Athens (in Greek). With permission.

environment. In this manner, about 35 different chemical elements, essential for the function of living cells, sequentially go through inorganic and organic forms and reversely. As a rule, these are used primarily by plants that find them in the environment as inorganic substances (Table 4.2) and next by animals and microorganisms. The latter decompose various waste, e.g. dead bodies and convert the waste into inorganic substances, gaseous or water soluble, which return to the ground, the water and the atmosphere.

It is observed that chemical substances necessary for metabolism are continuously being recycled, while energy is used by the ecosystems, then converted into heat and thus it is not reusable.

As a rule, the circulation of chemical substances gets beyond the limits of different ecosystems (Figure 4.4). The land and marine ecosystems are generally not closed but are interconnected, thus creating a planetary network of ecosystems, which develop on the ground, in the water and in the lower atmosphere levels around Earth's surface. This network constitutes the biosphere, within which the circulation of chemical elements is closed, creating the biogeochemical cycles of the various elements. This term means that recycling is realized inside the biotic (bio-) and abiotic (geo-) environments.

Various feedback and exchange mechanisms render the cycles self-regulating. A flow increase in one part of the cycle is quickly compensated by

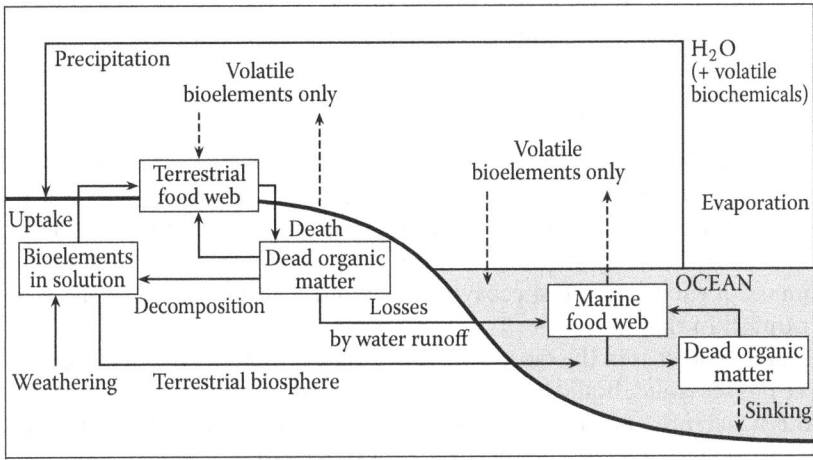

Figure 4.4 General schematic of nutrient cycling on a global scale. Movement of non-volatile elements, such as phosphorus, is largely one way, toward ocean sediments. (From De Angelis, D.L. (1992). *Dynamics of Nutrient Cycling and Food Webs*. Chapman & Hall, London; and Krebs, C.J. (2001). *Ecology*. Benjamin Cummings, San Francisco, CA. With kind permission from Springer Science + Business Media B.V.)

respective adjustments in other parts. Thus, the presence of cycles gives the biosphere a remarkable self-regulation capability. The result is the observed long-term stability of the various ecosystems, along with the continuous presence of essential substances in the various environments in generally stable analogies. Therefore, life and mainly producer organisms have the ability to obtain the substances they need from the space they live in.

The matter that moves in a biogeochemical cycle is often stored for a specific time frame in some abiotic or biotic component of the ecosystem, which is called a *reservoir*. The stored matter quantity is the storage, whereas the quantity that is transferred between reservoirs per unit time is the flow velocity. In an ecosystem that is in a state of equilibrium, $R_i = R_k$ applies, where R_i is the sum of inflow velocities and R_k is the sum of outflow velocities from one reservoir. Turnover time is the magnitude T, equal to M/R_i or M/R_k, where M = storage. T can be considered as the mean time that the matter remains within the reservoir. It serves as a sensitivity measure of a component of the ecosystem versus the variations caused by human activity in the biogeochemical cycles. Greater T periods indicate greater resistance in velocity flow variations. It has been estimated that time T in the reservoir "atmosphere" is for N_2 64×10^6 years, for O_2 7500 years and for CO_2 5 years.

Most important are the biogeochemical cycles of H_2O, C, O_2, N_2, P, and S. It should be noted that chemical element's circulation within the ecosystems, as well as energy flow, are phenomena that may be associated with pollution, e.g. toxic substance concentration in food chains.

4.3.1 Water Cycle

Water is the inorganic compound present in the greatest proportion within the living matter. Water percentage fluctuates from around 10% in plant seeds to 98% in some aquatic animals, whereas it is about 63% in the human body. The significance of water for life is evident by the fact that the biomass of each terrestrial ecosystem is almost proportional to the volume of rainfall in that area.

Water is present in the biosphere (atmosphere, hydrosphere, lithosphere) in three phases (solid, liquid, gaseous). Its distribution, however, is very unequal. If the water present in oceans could be uniformly spread upon the Earth's surface, a layer with a thickness of 2.7 kilometres would be created. If the same happened for the water of the glaciers, the layer would have a thickness of about 100 meters, whereas with underground fresh water the thickness would be 15 meters, for surface fresh water it would be 40 centimetres, and with the water-vapour present in the atmosphere only 3 centimetres.

The greatest part of the water cycle is realised between the sea and the atmosphere. Vapours liquefy in the clouds and descend on the Earth's surface in the form of rain, snow, hail, dew or frost. Water reaching land surface can follow three paths: penetration in the ground, evapotranspiration or surface flow. In the first case the containment and storage of water is ensured; vegetation and the soil's organic matter play a significant role in this process. In the second case, the water returns swiftly to the atmosphere, either through direct evaporation or through plant transpiration; plants accelerate to a great degree the return of the water to the atmosphere, since they absorb important water quantities through their roots and discard them into the atmosphere through their leaves. The third case is favoured by vegetation deficiency and constitutes the main factor of ground erosion. Ultimately, a large part of terrestrial waters end up in the oceans through surface or underground flow. Evaporation from the ocean's surface is the most important path water follows in order to return to the atmosphere.

4.3.2 Carbon Cycle

4.3.2.1 Main Features of the Cycle

Carbon, life's primary element, forms the most perfect of all biogeochemical cycles and circulates with a relatively rapid velocity in biocoenosis through food chains. The most essential inorganic forms of carbon presence in nature are carbon dioxide (CO_2) and calcium carbonate ($CaCO_3$). The part of carbon that is naturally circulating in the biosphere goes

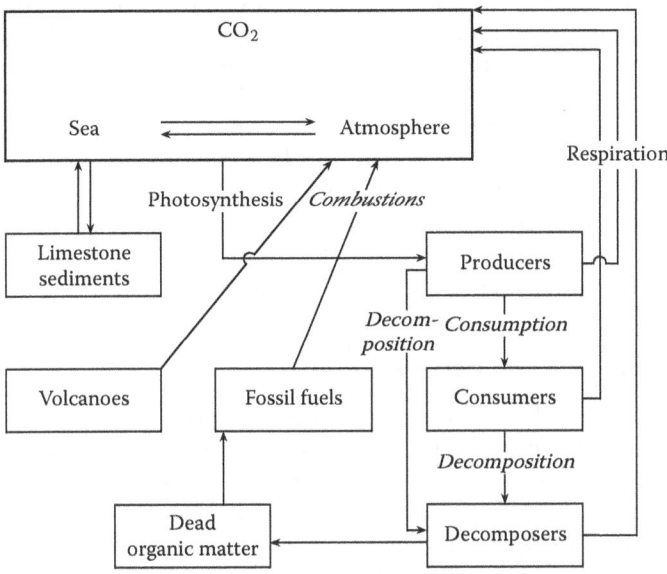

Figure 4.5 Carbon cycle. (Modified from Hadjibiros, K. (2007). *Ecology. Ecosystems and Environmental Protection,* 3rd edition. Symmetria, Athens (in Greek). With permission.)

through the exchanges between living organisms and CO_2. All land plant production is realised by CO_2 consumption through photosynthesis. This gas returns in the atmosphere through organic matter oxidation that takes place during the process of respiration.

The sea contains an important amount of dissolved CO_2, about 50 times greater than the one in the atmosphere. The dissolved gas allows algae and marine plants to photosynthesize, while at the same time it constitutes a reservoir which can absorb or render CO_2 to the atmosphere, thus playing a regulating role in the function of the whole system (Figure 4.5). The greatest quantity of dissolved CO_2 in marine ecosystems is found in the form of $(HCO_3)^-$. Generally, dissolved carbon in the water appears in the forms of CO_2, H_2CO_3, $(HCO_3)^-$, $(CO_3)^=$ which are in physicochemical equilibrium.

The normally rapid circulation of carbon among ecosystems, the atmosphere and the hydrosphere is, in many cases, decelerated. One of these cases is the accumulation of dead substances in the ground, where they create a dark-coloured mass that is called *humus* and plays a significant role in retaining water and nutrients. As a rule, humus conversion into inorganic materials with a simultaneous release of CO_2 takes place in a very slow manner, especially in the temperate and polar zones of the planet. Reversely, in tropical zones it is usually quicker, resulting in the usually slim thickness of tropical soils. Thus, when tropical soils are not protected by rich natural vegetation, they become infertile and susceptible to erosion.

The main reservoirs of the carbon cycle and the calculated storages are the atmosphere, the sea, living organisms, dead organic matter and fossil fuels. The carbon quantities stored in terrestrial living organisms' biomass, as well as those stored in the soil are of the same order of magnitude as the amount contained in the atmosphere; the carbon quantities stored in fossil fuels are estimated to be about 4–5 times greater. The greatest carbon quantities are accumulated in carbon rocks, mostly in $CaCO_3$ that contains a CO_2 quantity more than 40,000 times greater than that in the atmosphere.

4.3.2.2 Carbon Cycle Perturbations

The carbon cycle has a slow component. Through a series of chemical reactions and tectonic activity, carbon takes between 100 and 200 million years to move between rocks, soil, ocean and atmosphere. About 20% of carbon-containing rock contains carbon from dead organic matter that has been embedded in layers of mud. Heat and pressure compress the mud and carbon, forming sedimentary rock such as shale. In special cases, layers of organic matter become fossil fuel instead of sedimentary rock. During geological centuries, fossil coal, lignite, petroleum and natural gas were created by this process. It was a large-scale phenomenon during some of the geological periods, while at present it continues but is insignificant at a biosphere level. The carbon from sediments returns slowly to the atmosphere through volcanoes. By burning fossil fuels, human populations emit per year 100–300 times more carbon dioxide than the volcanoes.

Before the industrial era, carbon's fast recycling in ecosystems has been almost perfect and stable. However, in previous geological periods, photosynthesis exceeded respiration in the biosphere, resulting in a decrease of CO_2 and an increase of O_2 in the atmosphere, and the storage of carbon in the soil and sediments. Of great importance is also the phenomenon of dissolved in water CO_2 conversion to $CaCO_3$. It is partly due to chemical precipitation, but mostly due to the biological activity of small marine animals that construct their skeletons with calcium carbonate which, after their death, is deposited in the bottom's sediment. Thus, huge sediment layers of calcium carbonate were created during various geological periods.

The importance of the perturbations of carbon cycle's fast component becomes apparent when the history of Earth's atmosphere is studied. The atmosphere was not always as it is at present. Initially, large amounts of hydrogen and light inactive gases were present, which were soon lost, that is, around 4–5 billion years ago. Next, a reductive atmosphere was created, containing no oxygen, but mostly gases such as nitrogen, methane, ammonia, water vapour and carbon dioxide. Some hundred million years after the creation of life, oxygen began being produced via the process of photosynthesis. Through various chemical reactions, the atmosphere transformed into its present composition, which mainly contains nitrogen, oxygen and small quantities of argon, water vapours and carbon dioxide. A decisive parameter in the development and propagation of living organisms was the ozone layer in the stratosphere; it was created when oxygen reached an important concentration in the atmosphere. The ozone layer began to protect life against the sun's ultraviolet radiation (Chapter 9). Thanks to that protection, living organisms were able to survive out of the water, plenty of forests grew on land and photosynthesis dominated. It is estimated that the total amount of oxygen produced by plants during Earth's history is tenfold the amount present in the atmosphere today. Therefore, the corresponding amount of carbon dioxide has been removed from the atmosphere through photosynthesis.

With the stabilisation of the carbon cycle, the atmosphere's composition became approximately similar to the present one and has since been stable. For millions of years, stability prevailed, with plant production consuming, almost completely, all living organism's respiration. However, as early as the 19th century, this balance had been upset. Extensive deforestation and use of fossil fuels have released large amounts of carbon into the atmosphere, in a CO_2 form; these had remained stored in forest biomass or fuels respectively. At present, humans, with the extensive use of fossil fuels reserves, consume in a year as much O_2 as photosynthesis produced in 1000 years. Without human interference, the carbon in fossil fuels would leak slowly into the atmosphere through volcanic activity over millions of years. Humans accelerate the process, releasing carbon that took millions of years to accumulate. In 2009, humans released about 8.4 billion tons of carbon into the atmosphere. This reversal, though, of natural processes does not create a threat of O_2 deficiency, since its quantities in the atmosphere are quite large and its consumption negligible compared to them (Chapter 4, Section 4.3.4). However, CO_2 excess poses risks, since its presence in the atmosphere is small and its percentage has been significantly increased (Chapter 9). It is estimated that atmospheric air's CO_2 was in an analogy of 290 ppm in the middle of the 19th century; it has reached 415 ppm in 2021.

4.3.3 Nitrogen Cycle

Nitrogen is necessary for the creation of proteins. Despite its abundant presence in the atmosphere in its molecular form, it is chemically inactive and cannot be easily broken down by organisms. Most producer organisms can only absorb the nitrogen in the forms of ammonium nitrate $(NH_4)^+$ and nitrogen nitrate $(NO3)^-$. Thus, even though atmospheric N_2 constitutes a huge storage, its cycle, and along with it the life process in our planet, are limited by the conversion speed of atmospheric N_2 to its usable bound forms.

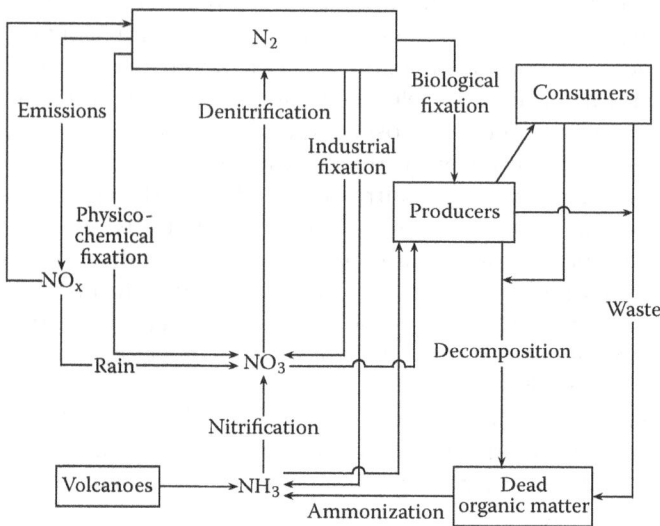

Figure 4.6 Nitrogen cycle. (Modified from Hadjibiros, K. (2007). *Ecology. Ecosystems and Environmental Protection*, 3rd edition. Symmetria, Athens (in Greek). With permission.)

Producer organisms integrate nitrogen in the prepared proteins. The rest of the organisms supply themselves with the necessary organic nitrogen from producer organisms, whereas some specialised decomposers break down nitrogen compounds of dead organic matter. The greatest part of nitrogen remains in the ground or water in the form of nitrate or ammonium salts and is reused by plants. A smaller part is released as gaseous N_2 in the atmosphere and completes the cycle. Finally, some N_2 quantities get out of the cycle, since they end up as deposits in the bottom of the oceans. A small enrichment of the atmosphere with N_2 occurs from volcanic activity. The most important lines of the nitrogen cycle appear in Figure 4.6. The main flow lines of N_2 are fixation, nitrification, ammonisation and denitrification. Fixation is the conversion of molecular N_2 into the usable forms of ammonium and nitrates.

Natural fixation is realised either through physicochemical processes, mainly by the influence of electrical discharges in storms, or by various biological processes. According to estimations, annual biological fixation is 54×10^6 tons of atmospheric N_2, and the respective physicochemical one is 7.6×10^6 tons. Other estimations raise the biological fixation to 150×10^6 tons.

Biological fixation is realized by heterotrophic organisms, as well as by some autotrophic-photosynthetic organisms. A characteristic example is that of rhizobium, which develops in the roots of leguminous plants, such as clover, beans or peas. This bacterium fixes N_2 which will be used by the leguminous plants, while it takes from these plants the necessary energy and substances (symbiosis). Additionally, non-symbiotic bacteria exist which are able to fix N_2, such as the aerobic nitrobacterium and the anaerobic clostridium. Some algae that belong to the blue-green algae group (cyanophyta) can also realize N_2 fixation.

Artificial nitrogen fixation is a 20th century invention and is mostly realised through industrial production of artificial fertilizers, which is of the same order of magnitude with biological fixation. In 1968, industrial N_2 fixation was equal to 30×10^6 tons. Significant artificial N_2 fixation is also caused as a side effect of internal combustion engines functioning or of some industries which emit nitrogen oxides (NO_x) in the atmosphere. These oxides represent, on one hand, a serious air pollutant (Chapter 9), and on the other hand, they are oxidised to nitrates and end up either in waters or the soil.

Nitrogen incorporated into living organisms returns to the environment in an inorganic form, ammonia (NH_3), or the product of its ionisation (NH_4)$^+$, as a result of dead organisms' decomposition and their waste. This phenomenon is defined as ammonisation.

The chemoautotrophic bacteria nitrosomonas and nitrobacterium are interesting for pollution control technology. They are strictly aerobic bacteria and obtain the necessary energy by oxidising ammonia to nitrites (NO_2)$^-$ and nitrate (NO_3)$^-$, according to the following chemical reactions:

$$NH_3 + 3/2\,O_2 \rightarrow NO_2^- + H^+ + H_2O + energy$$

$$NO_2^- + 1/2\,O_2 \rightarrow NO_3^- + energy$$

The phenomenon of biological conversion of ammonia to nitrates is defined as nitrification. Nitrates are, in general, more conducive to biological processes but also constitute a pollutant of aquatic bodies. Nitrification is a water deoxygenation factor, due to aerobic oxidation. There are also some other bacteria categories, which can convert nitrates to nitrites and nitrites to ammonium nitrate.

Denitrification is the process of nitrate reduction to N_2 and N_2O, during anaerobic oxidation of organic matter by microorganisms. N_2O is reduced to N_2 by further bacterial action or by photochemical reactions within the atmosphere. Thus, N_2 returns to the atmosphere, closing the cycle. Without denitrification, all the nitrogen in the atmosphere would have been bound, during Earth's long history, in the land and oceans in the form of nitrates.

The main reservoirs of the nitrogen cycle and the estimated storages is the sea (dissolved, $20,000 \times 10^9$ tons), or the atmosphere ($3,800,000 \times 10^9$ tons), organisms (13×10^9 tons), dead organic matter (1660×10^9 tons), and depositions ($4,000,000 \times 10^9$ tons).

Nitrogen cycle perturbation by man and the resulting nitrogen accumulation in the biosphere have the following environmental consequences:

- Eutrophication of aquatic ecosystems in which nitrogen is a limiting factor (Chapter 8)
- Burdening of aquatic ecosystems (rivers, lakes, oceans, groundwater) with nitrates or ammonia (Chapter 8)
- Burdening of the atmosphere with nitrogen oxides, which are toxic, contribute to the creation of acid rain and intensifie the greenhouse effect (Chapter 9)

The problem, summarily, is that the pace of total nitrogen fixation is at present significantly greater than the pace of total denitrification. Therefore, the soil and the waters, at least on a local scale, accumulate each year a surplus of fixed nitrogen. Dealing with this perturbation requires the restoration of the nitrogen cycle's more natural function, i.e. the reduction of industrial fixation, nitrogen oxide emissions in the atmosphere, and generally, of every kind of anthropogenic nitrogen fixation, as well as the protection or enforcement of natural and the development of artificial denitrification processes, such as removal of nitrogen from wastewater by tertiary treatment (Chapter 10), treatment of car air emissions through catalytic converters (Chapter 10) and creation of artificial wetlands.

4.3.4 Oxygen Cycle

Oxygen is the most abundant element both in Earth's crust and in the living matter; it is present in numerous chemical compounds. Oxygen is necessary to life, since it allows for aerobic respiration to occur but in great concentrations it can be toxic for organisms; it can also cause spontaneous ignition of dry vegetation or biomass. Its cycle is complicated and creates special sectors in the atmosphere, hydrosphere and lithosphere. However, it is realised for the largest part between the atmosphere and the biosphere, through photosynthesis which produces O_2 and respiration which consumes it. Thus it portrays essentially the opposite image of the carbon cycle, since the movements of the one are executed in the opposite direction of those of the other. Due to the existence of a large O_2 quantity in the atmosphere, its total recycling by the ecosystems takes thousands of years.

Atmospheric O_2 as well as O_2 compounds of many surface rocks (limestone, iron oxides etc.) have a biogenetic origin. O_2 did not exist in the primitive atmosphere and was produced by the photosynthesis of plants and other organisms. Excessive O_2 production ceased millions of years ago, since photosynthesis and total respiration reached an equilibrium, resulting in the end of O_2 accumulation in the atmosphere.

Human action interferes in the O_2 cycle through the destruction of natural vegetation and the use of fossil fuels. This interference significantly affects the carbon cycle that is more sensitive to perturbations because the quantity of CO_2 in the atmosphere is small; it does not perturb the cycle of O_2 because it is abundant (21%) in the atmosphere. However, perturbations take place in some parts of the O_2 cycle, with the most important being the depletion of the ozone layer (Chapter 9). Other perturbations are the problem of deoxidation, occurring when O_2 does not have an adequate concentration in the environment, thus becoming a limiting factor for some populations. Such phenomena often arise in aquatic environments (Chapter 8), though not in terrestrial environments, since even in a polluted urban atmosphere (Chapter 6) O_2 has a stable concentration.

4.3.5 Phosphorous Cycle

Phosphorous (P) is an important element of living matter and its relative scarcity often turns it into a limiting factor for organisms' development. Many ecosystems' primary production might be finally controlled by phosphorous, thus this element is considered to be the ultimate regulator of the other biogeochemical cycles. Many scientists consider P as the weak link in life's chain on Earth, since its cycle is relatively imperfect and open.

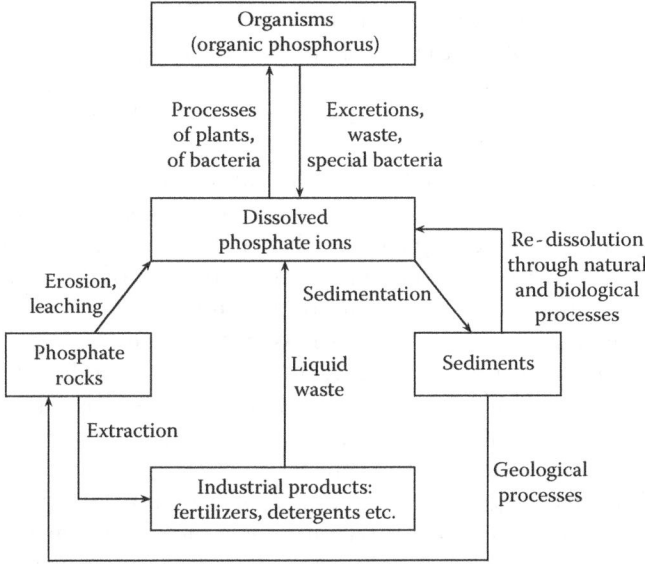

Figure 4.7 The natural and anthropogenic phosphorus cycle. (From Hadjibiros, K. (2007). *Ecology. Ecosystems and Environmental Protection,* 3rd edition. Symmetria, Athens (in Greek). With permission.)

The total phosphorous reservoir is found in the lithosphere and enters the biosphere's circulation through the rinsing of the ground surface (Figure 4.7). Through water flow it reaches terrestrial and aquatic ecosystems and is absorbed by plants and then goes through food chains. As in the other cycles, various decomposers in the soil or water break down dead organic matter and produce inorganic P, available to plants; P is utilized by producer organisms in the form of soluble orthophosphates. Insoluble compound conversion to orthophosphates is performed by specialized microorganisms. Dead organisms' decomposition and their waste yield P to the environment in the form of orthophosphates, which, however, can easily form insoluble compounds with calcium and aluminium ions, which are quite common in the soil.

Recycling of P is almost complete in terrestrial ecosystems, where dead animals' skeletons decompose and the contained P again becomes exploitable. The same occurs in fresh waters or shallow seas, where bound P deposition can easily return to food chains. Conversely, in the open ocean that represents 85% of the marine surface, a continuous loss of P occurs due to the accumulation of organic matter and mainly dead bodies, bones etc. in the abysses and generally in great depths. Ocean upward currents enriching the shallow water around ocean shelves with nutrients cannot restore the deep deposits in P's biogeochemical cycle. Thus, phosphoric compounds deposited in the bottom of the ocean are retracted from the biosphere. Certainly their recycling is still possible through tectonic upward movements of the sediments, but in a geological scale, i.e. in dozens or hundreds of million years. Thus, the P cycle remains partly and for the short term open.

A significant part of P returns from the sea to the land through seabirds, which consume fish and leave their waste on land. Nevertheless, these movements of P are insufficient in compensating for the outflows from the land to the sea.

Human activity aggravates P cycle's natural imperfectness. Large quantities of phosphoric substances excavated from the lithosphere's natural deposits are transformed into phosphates by the fertilizer or detergent industries, ending up in the hydrosphere and partly at the bottom of the oceans.

Phosphorus appears to be the limiting factor in more cases than any other element. One reason is the easiness with which it creates insoluble compounds; this fact inhibits its reception from plants and its transportation by surface and underground water. Another reason is P's inability to create gaseous compounds in the natural environment, which deprives it of the ability to circulate in the atmosphere.

Urban fluid discharges contain significant P quantities, mainly due to detergent usage, and the same is true for some industrial waste. The presence of P in liquid waste has become the reason for far gone eutrophication states in lakes and coastal water, where it often is the limiting factor. The potential contribution of phosphate fertilizers to eutrophication is smaller due to the insoluble compounds that P creates with elements of the soil.

4.3.6 Sulphur Cycle

Sulphur (S) is an essential component of some amino acids. Its cycle is mostly realised in the aquatic environment and the soil, whereas gaseous sulphur compounds, such as sulphur dioxide and hydrogen sulphide, have a secondary part in its circulation in the biosphere. Sulphur's basic source for organisms is sulphide salts that are dissolved in the water. These are absorbed by plants and are used for protein production. Sulphur's organic compounds circulate like the other substances through food chains and are finally decomposed by special bacteria that generate hydrogen sulphide. These microorganisms live in anaerobic conditions within the black silt existing in the bottom of certain seas or lakes; many of them produce the energy they need by the reduction of sulphide compounds. Conversely, some autotrophic bacteria construct organic substances thanks to the energy they produce from the oxidation of hydrogen sulphite to sulphur compounds. Additionally, a part of S found in the bottom of the sea settles in the sediments in the form of sulphide iron, whereas phosphorus is simultaneously released from the sediments, becoming soluble in the water. This is a remarkable example of the combination and mutual regulation of two different biogeochemical cycles. Finally, the biosphere is constantly enriched by gaseous S compounds through volcanic activity, as well as by human actions, as the combustion of fossil fuels, which have become significant, especially in the last century.

4.4 ECOSYSTEMS' CHANGE IN TIME

4.4.1 Ecological Succession

Ecosystems are continuously changing under the influence of external and internal factors. The time scale of the short-term change is that of the organisms' life span order. Long term, ecosystems' changes correspond to scales of hundreds, thousands and millions of years, following the changes of geological conditions or of the climate, as well as the species' evolution. Moreover, a middle term change is observed in certain ecosystems in the form of gradual maturation of biological communities. It is completed in time scales of years, decades or centuries, at the most.

A piece of land that came to the surface by volcanic action or by the recess of a glacier does not initially have traces of life. Gradually, as the rock is exposed to the action of the wind, of the water and temperature fluctuations, it will suffer erosion. Sooner or later, the first plants and microorganisms will invade the area and they will accelerate the formation of soil. The invasion of many plant species will follow, which will create more soil through their roots and their waste. With the establishment of a rich plant life, herbivorous animals will invade the area and the carnivores will follow, until a complete ecosystem is formed.

Phenomena of successive biological communities' creation are called ecological succession. The succession continues until a mature and relatively stable (climax) biocoenosis is formed, which is characteristic of the specific climate and soil of the area. During the succession, at least in the initial stages, species diversity increases and the trophic dependence among species tends to form a complex network.

We can distinguish between primary and secondary succession. The first begins in places where ecological functions develop on a previously abiotic environment. The second takes place when an existing biocoenosis has been degraded or destroyed and is gradually recovering through successive pioneer biocommunities. A classic example of primary succession is ecosystem creation in a new volcanic island. Another example is the phenomenon where natural sand dunes act as barriers and create shallow coastal lakes that may gradually develop into forests. Secondary succession is observed when a forest burns down and the natural vegetation is regenerated and forms successive phytocoenoses that ultimately become a mature forest. Another example is the development of a new biocoenosis where the waste of various organisms has contributed to the creation of soil.

4.4.2 Complexity and Stability of Ecosystems

One of the theoretical problems of ecology is that at ecosystems' level, persistence is observed, whereas at lower levels, such as populations, for example, there is no steadiness. Linking ecosystem's persistence with diversity or with the complexity of the ecosystem's trophic networks has been suggested; this idea is only partially accepted and the question remains open.

A system's persistence means that the system permanently remains more or less in the same state, which is connected with stability or resilience, i.e. its endurance against perturbations. Cohesion of a system is the total of the relations that connect its parts. The complexity of a system corresponds to the number of sub-systems that can co-operate for some goal, whereas diversity corresponds to the number of sub-systems that can act independently. In the case of ecosystems, diversity can be conceived as the number of the ecosystem's different species or as a mathematical expression that depends on that number. It is obvious that a greater species' diversity means a greater possibility for the existence of multiple networks of relationships and therefore of greater complexity as well. In this case, a consumer organism's population can use other populations for food and the system eventually has increased ability for resistance to perturbations.

There must be a distinction between the ecosystem's ability to be stable in the face of small perturbations and its corresponding ability (resilience) to return to its previous state after great perturbations. Great complexity may help the ecosystem to overcome relatively small perturbations in a generally predictable environment. Complexity, however, is a fragile state that becomes possible mainly because of the environment's stagnation. Complex natural systems tend towards simplification when confronted with perturbations that surpass their normal experience. A non-predictable environment requires a rather simple and robust ecosystem.

The connection of diversity or complexity with the stability of ecosystems has been discussed for many years. The thesis of connection has been based on various observations and arguments, summarized as follows:

- The simple mathematical models of predator-prey end up in non-stable systems.
- Laboratory experiments with simple predator-prey systems usually end up with the extinction of the first or both species.
- Small islands' ecosystems (that are relatively simple) exhibit small resistance to the introduction of new species.
- Population explosions happen more often in single-crop farming and less often in more complex ecosystems, whereas no population explosions are observed in tropical forests.
- Parasites create more serious problems to cultivations where many or strong pesticides have been applied.
- Ecosystems in northern and arctic areas exhibit great oscillations (possibly because their complexity is not great enough to eradicate them).

Theoretical and experimental ecologists, however, have developed many counter-arguments that can be summarized as follows:

- The simple mathematical models of population dynamics are not stable, but the more complex ones are even less stable.
- There are natural "single-crop" systems (e.g. marshy coastal vegetation) that are very stable.
- The most significant parasites that present population explosions (locusts, rodents etc.) have a great variety of enemies.
- Islands do not exhibit lack of stability; they are simply more vulnerable. There are complex systems as well that are vulnerable (i.e. forest destruction due to the chestnut's parasite).
- The stability of the coral reefs' complex ecosystem (aquatic equivalent to the tropical forest) was hard hit by the invasion of a starfish species.
- Natural ecosystems, complex or not, are products of a long-time co-evolution of their populations. Evolution managed to find the, perhaps small, parameters' space that give long-term stability to the system. Thus the comparison of a tropical forest with an agricultural system is not relevant, because a natural system, which is a product of a long evolution, is compared with an artificial one.

The subject remains open, but in general, the relationship between ecosystems' complexity or diversity and stability does not apply. For the time being, there is no way to pre-estimate the consequences of a species' extinction from a biocoenosis with satisfactory accuracy. In some cases, such change has perceptible results but in others it does not. It is reported that in the case of a marine community with 15 invertebrate species, the removal of one caused the reduction of the other species to 8.

The relation between diversity and stability has been linked to the problem of environmental protection against human activities. It is often suggested that strict measures should be taken for the species' diversity conservation, based on the argument that otherwise, the general equilibrium of ecosystems is at risk. This argument does not have strong documentation. However, as a rule, the reduction of biodiversity brings about adverse consequences to ecosystems and man and it should be avoided, if only for preventive reasons.

The concept of species diversity in a biocoenosis is related with the richness of its functions. Diversity could firstly be defined as the multitude of species in the biocoenosis. The richness of functions, though, is related not only with the multitude of species but with their relative abundance as well. Let us suppose that there are two communities, A and B, with the same total number of individuals $N = 100$ and the same species, 4 for each community. Let us also suppose that the population sizes in community A are 25, 25, 25, 25, and in community B they are 94, 2, 2, 2. Although both communities have the same multitude of species, one can expect that A will be characterized by a greater richness of functions compared to B. This is why diversity indices that constitute a measure of the relative species abundance have been proposed. One of those indices that was proposed by Simpson is the following:

$$\text{Diversity index: } DI = \frac{N(N-1)}{\Sigma_i n_i (n_i - 1)}$$

where
  N = total number of community's individuals
  n_i = number of individuals of species i

The diversity index, as defined by the above relationship, is the multitude of random couples that should be taken from the whole population of the community, so that there is a probability of at least 50% that a couple composed by individuals that belong to the same species will be obtained. If the formula for communities A and B is applied, we will have $(DI)_A = 4.125$ and $(DI)_B = 1.132$.

4.5 FUNCTIONS OF AQUATIC ECOSYSTEMS

Aquatic ecosystems are distinguished into the marine ones, which include open and closed seas, and the freshwater ecosystems, which include lakes, rivers, springs and most of the wetlands. A special case of marine ecosystems are coastal aquatic ecosystems; they generally are wetlands. Underground waters are not considered to be autonomous ecosystems. The study of aquatic ecosystems is especially important for the thorough examination of most of the biogeochemical cycles as well as the various water pollution phenomena (Chapter 8).

4.5.1 Trophic and Other Characteristics of Aquatic Ecosystems

The main producer organisms in aquatic ecosystems are algae (phytoplankton); plants with roots are confined in the shallow areas, and for this reason usually play a secondary role in photosynthesis and therefore in the trophic chains of aquatic ecosystems; the contribution of some photosynthetic cyanobacteria is generally marginal.

Consumer organisms are usually ranked in 3–4 trophic levels. Primary consumers include a great variety of organisms, such as protozoans, jellyfish, molluscs, worms as well as some small fish. The rest of the consumers are mainly bigger fish, as well as some mammals and birds. The category of decomposer organisms includes bacteria and fungi.

Aquatic ecosystems' organisms are distinguished in plankton, nekton and benthos. Plankton is the total of organisms that float, with a mainly passive movement caused by waves and currents. It consists of algae (phytoplankton), herbivorous animals (zooplankton) and decomposers. Nekton consists of organisms that have an active movement in the water. Benthos are organisms that live in the bottom of the aquatic space and seek their food in silt.

Contrary to many terrestrial ecosystems, the availability of nutrients usually determines the aquatic ecosystems' productivity. The concentration of nutrient salts is generally low in not polluted water ecosystems. Typically, the concentration of nitrates in sea water is only 1/10,000 of their concentration in fertilized soil. For this reason, when pollution adds nutrients to the water bodies, a dramatic production increase is observed. The phenomenon is called eutrophication and shows the significance of nutrients as a limiting factor of aquatic primary production. Since the atmosphere transfers CO_2 to the aquatic ecosystems, carbon as a rule does not constitute a limiting factor for eutrophication. This role is usually claimed by nitrogen or phosphorus (Chapter 8). Untreated liquid waste and surface runoffs from fertilized fields usually contain nitrates or phosphates. They often cause pollution to the corresponding water receptors, especially in lakes and closed marine bays. The phenomena of materials' transfer from the terrestrial to the aquatic environment and gases' exchange between the atmosphere and water are significant for the aquatic ecosystems' function.

The availability of basic nutrients is an especially important factor for lakes; thus, limnologists classify them into the categories of oligotrophic and eutrophic lakes. In the geological and ecological history of lakes, there is a general tendency from oligotrophism to eutrophism. As centuries go by, debris are deposited in the lakes and, according to the nature of the parent rock, nutrients are continuously accumulated and this favours an increasing primary production. This natural course has often been accelerated by man through wastewater disposal as well as the wide use of fertilizers.

Oxygen is also a significant factor of aquatic ecosystems' function. The maximum concentration in which it may be diluted in the water is low, one milligram per litre, while the speed by which it is absorbed by the atmosphere is also small. A usually secondary oxygen contribution comes from the aquatic producer organisms. Algae, as well as aquatic plants, enrich the water with oxygen through photosynthesis, while they consume a smaller quantity of oxygen through respiration. When dissolved oxygen is absent, aerobic respiration is not possible, and therefore the aquatic environment only can support lower forms of heterotrophic organisms, basically anaerobic bacteria. Thus O_2 can be the limiting factor for many aquatic populations, because of aerobic respiration.

4.5.2 Temperature Layering and Seasonal Inversions

In the summer, the phenomenon of temperature layering or stratification is observed in deep lakes and seas (Figures 4.8 and 4.9). A surface layer of water, a few meters thick, acquires a relatively high and uniform temperature. This layer is called *epilimnion*. It is characterized by an almost stable density and its masses move easily by the wind and transport currents. Thus the epilimnion tends to maintain a high degree of mixture.

Under the epilimnion we find the thermocline or mesolimnion, a layer of water of a relatively small thickness, characterized by a rapid decrease of temperature with the increase of depth.

The water layer found under the thermocline reaches as far as the bottom; it is called *hypolimnion* and is characterized by a lower temperature that slowly decreases as it goes deeper and it can get as low as 4°C, that is, the temperature of maximum density. The horizontal movements in the hypolimnion are weak, while the vertical are almost non-existent.

In the fall, the surface water masses get colder, and they become more dense and sink, whereas warmer and thinner water masses rise to the surface. This movement of water masses is called *autumn* or *great inversion* and tends to establish a stable temperature in the whole depth of the water.

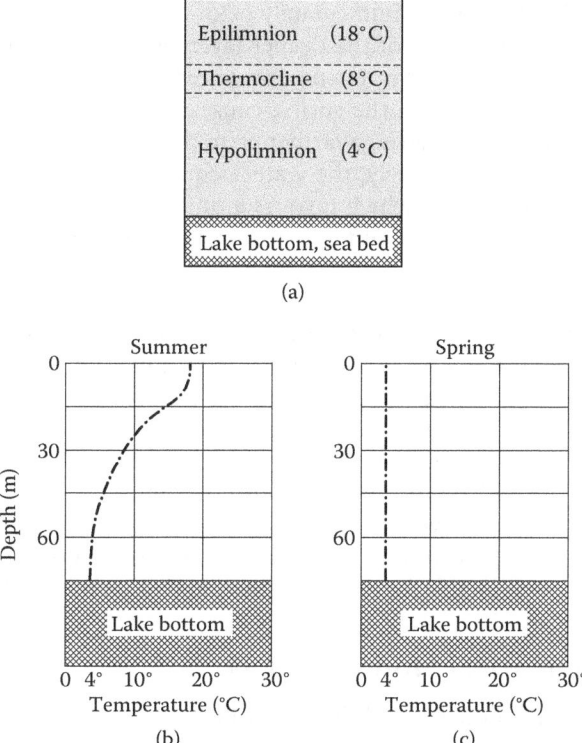

Figure 4.8 Temperature layering of a temperate zone's deep lake. (Modified from Hadjibiros, K. (2007). *Ecology. Ecosystems and Environmental Protection*, 3rd edition. Symmetria, Athens (in Greek). With permission.)

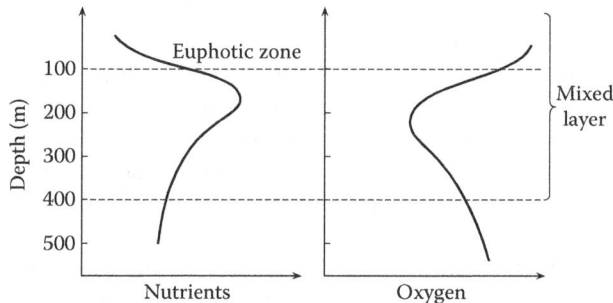

Figure 4.9 Oxygen and nutrient profiles in the ocean. Concentration of ambient nutrients and oxygen through the mixed layer are inversely correlated, reflecting uptake and production by photosynthetic organisms. (From Colinvaux, P. (1993). *Ecology 2*. Wiley, New York. With kind permission from John Wiley & Sons.)

In colder climates during the winter, surface water temperature can get as low as almost 0 °C (e.g. water immediately below a surface ice layer), whereas in this case, the temperature of the deeper layers is near 4 °C. In this way, a winter layering is created that is much less marked and the reverse of the summer one. Temperature rise in the spring causes a smaller, spring inversion, given that surface masses, getting warmer up to the temperature of maximum density, tend to sink. Ultimately, the water acquires a uniform temperature and density in all its depth, which favours a quite strong circulation.

The ecological significance of temperature layerings and seasonal inversions is great, since:

- Temperature changes directly affect organisms' activities.
- Summer layering obstructs mixing; consequently the oxygen that is absorbed by the atmosphere basically circulates only in the epilimnion, whereas the hypolimnion tends to be deprived of oxygen. During autumn inversion, though, the hypolimnion's masses come to the surface and become enriched with oxygen.
- During seasonal inversions, nutrients get mixed and redistributed. Nutrients that had sunk to low water masses rise to the surface where they are now available for photosynthesis and other metabolic processes.

4.5.3 Mathematical Model of a Lake

The mathematical model of a lake that follows is simplified and aims at aiding the understanding of some main function mechanisms of aquatic ecosystems and showing the general principles of mathematical simulation of those mechanisms.

A non-stratified lake of a volume V is examined, with the assumption of complete mixture, i.e. with the assumption that nutrients and organisms are uniformly scattered in volume V. Therefore the study can be conducted on the basis of concentrations (mass/volume) of organisms and nutrients that are spatially stable. It is also accepted that the limiting factor of the ecosystem is phosphorus. Carbon, nitrogen and the other nutrients, as well as solar energy, are abundant for the whole duration of the phenomenon under study (e.g. from May to September). The course of the phenomenon is examined in temporal steps of 24 hours, on the basis of the average 24 hours solar radiation.

In reality, when the organisms die they sink to the bottom, decompose and supply again the water body with phosphorus and other nutrients. This recycling of phosphorus is not taken under consideration.

During the time $t = 0$, the concentrations of inorganic phosphorus, producer organisms, and consumer organisms in the lake are C_0, X_{p0}, X_{c0} correspondingly. In the time $t = 0$, inflow of inorganic phosphorus in the lake begins, with a speed (mass/volume/time) that equals L. Speed L is a known deterministic or stochastic function of time. We assume that the incoming phosphorus scatters instantly and uniformly in the lake (complete mixture) and remains diluted in the water in its totality without any observed phenomena of sinking in the bottom of the lake.

Organisms are divided into only two categories: producers and consumers. State variables of the ecosystem are the concentrations of organisms X_p and X_c and the concentration of inorganic phosphorus C. For each one of the two categories we assume that the biomass creation speed is analogous to the consumption speed of the limiting food, with corresponding analogy coefficient Y_p and Y_c. We accept that the food consumption speed R is described by the following empirical relation (Monod equation):

$$R = \frac{KF}{K' + F} \times (\text{mass/volume/time})$$

where F and X are the concentrations of food and organisms in time t, while K and K' are constants. Food for producer organisms is inorganic phosphorus, while for consumers it is producer organisms.

We accept that the speed of decay (death) per organisms' mass unit is stable in time and equal to b_p for producer organisms and b_c for consumers. Based on the above, three equations of masses' (concentrations') change in time for the lake's volume unit can be written: one for the mass of consumer organisms (X_c), one for the mass of producers (X_p), and one for the mass of inorganic phosphorus (C).

Consumer organisms

Change of mass = (speed of creation – speed of decay) dt.

Therefore:

$$dX_c = \left(Y_c \frac{K_c X_p}{K'_c + X_p} X_c - b_c X_c \right) dt$$

Producer organisms

Change of mass = (speed of creation – speed of decay – speed of their consumption by consumer organisms) dt.

Therefore:

$$dX_p = \left(Y_c \frac{K_p C}{K'_p + C} X_p - b_p X_p - \frac{K_c X_p}{K'_c + X_p} X_c \right) dt$$

Phosphorus

Change of mass = (speed of entrance − speed of its consumption
by producer organisms) dt.

Therefore:

$$dC = \left(L - \frac{K_p C}{K'_p + C} X_p \right) dt$$

If we know the constants Y_c, K_c, K'_c, b_c, Y_p, K_p, K'_p, b_p, the initial concentrations X_{c0}, X_{p0}, C_0 and the function $L(t)$, we can arithmetically solve the system of differential equations step by step (e.g. for temporal step Δt = 24 hours) and find the values of state variables X_c, X_p, and C for successive times t.

Part 2

Fundamentals of Environmental Problems

Chapter 5

Values, Alterations and Degradation of the Natural Landscape

DOI: 10.1201/b14609-5

5.1 MAN'S HISTORICAL INFLUENCE ON THE BIOSPHERE

In contrast with the rest of the biological populations of the planet, man (*Homo sapiens*) has based his survival rather on the modification of the environment than on his adaptation to it. Human populations have brought about significant changes to the natural landscape with their activities; they have caused the extinction of many plant and animal species and major disturbances to the biosphere functions.

The civilizations that flourished in various places and eras found themselves in greater or smaller competition with the natural environment. Some of them maintained a partially harmonious coexistence with nature, e.g. the American Indians, whereas the civilizations that spread and predominated par excellence, especially European civilization, operated on the basis of the subjugation of the rest of nature to human desires. The demand for continuously greater security and comfort on the part of man caused the phenomena of multiform environmental modification in the various eras. The adverse side effects, which were an inextricable consequence of that process, gradually acquired grave dimensions, especially during the last two centuries, with the serious degradation of extended areas and the emergence of great environmental problems at a global level.

Biosphere is the planetary ecosystem, and it exhibits great complexity in relation to its structure and function. It has exceptional diversity in relation to both the physical and chemical substratum and the biological content, since it contains tens of millions of flora and fauna species that come from many more species that lived in the recent and distant past. This system is distinguished for its remarkable stamina in disturbances. Its stability successfully resisted different geological or physicochemical factors during the geological ages, until the emergence of man.

By the prehistoric era, human activity had already been a source of disturbance for the biosphere and caused continuous degradation that was first manifested by the decrease of natural vegetation and biological diversity. Man's impact began to be significant when fire, man's first technological achievement, was discovered. Palaeolithic hunters began to exert intense pressure on the natural ecosystems. Hundreds of thousands of years ago, the use of fire, with the purpose of scaring off or trapping prey, laid bare the primary forests in the areas where human populations developed. Later, fire and axe became the tools for extended deforestation by nomad shepherds with the purpose of creating pastures or expansion of open areas where the wild herbivorous animals develop better. Those activities coincide with a significant decrease of the wild fauna in the corresponding areas, and many species, especially large mammals and birds, disappeared, hunted by Palaeolithic man.

Systematic intervention to the natural environment began during the Neolithic era, 8000–10,000 years ago, when man settled in a certain location and started agricultural exploitation. The successive technological inventions led to the agricultural revolution. The invention of pottery rendered product storing possible, and at the same time paved the road for metallurgic techniques. Domestication of certain herbivorous animals enabled the development of livestock farming and necessitated the conversion of large stretches of land to pastures. The development of agriculture and livestock farming led to the extinction of lowland forests in almost all the regions of Earth, because they were turned to agricultural areas. The thoughtless exploitation of the lands brought about erosion and desertification in many cases. The increase of available food quantities enabled a significant demographic increase, which in turn caused further expansion of cultivations and pastures to the detriment of the natural environment. The necessary land became available through the natural ecosystems retreat and the substitution of primary vegetation in very large areas.

As centuries went by, the performance of agricultural techniques improved; as a result, the necessary for nutrition per capita land surface decreased. However, agricultural production was expanding in every available space due to various historical developments, mainly because of the human population increase. The cultivation of marginal lands, deforestation of areas with a marked relief or vulnerable soil, unsuitable irrigation of friable earth, over-grazing of the natural vegetation and use of fire were widely applied practices that had disastrous consequences. Even traditional shifting agriculture caused serious degradation, transforming forests to savannahs in many areas.

The environmental impacts of the agricultural civilization were significant but did not cause major disturbances to biosphere function. Natural cycles and the energy flow in the biosphere were not substantially affected; thus the self-regulating capability of the global environment remained intact. On the contrary, industrial civilization disturbed the functions of the biosphere to a great extent and led to destabilization conditions. The course that began from the end of the 18th century, took shape during the 19th century and reached its peak during the 20th century was associated with the gradual perturbation of the biochemical cycles and the significant degradation of the productive capacity of certain natural resources. Energy consumption, a symbol of the industrial era, constituted the main perturbations' source, and in combination with the new massive production processes, caused the accumulation of all kinds of waste and the pollution phenomenon at a global level. At the same time, alterations of the natural environment and over exploitation of natural resources (e.g. over fishing) that had begun in previous periods were intensified, resulting in accelerated decrease and degradation of biodiversity, of the forest cover of many terrestrial and marine ecosystems and of natural landscapes and seascapes.

By the late 20th century, the beginning of a meta-industrial era that came dimly into sight with the new millennium gradually created a new frame for human activities, where the dilemma between development and environmental protection came to the foreground (Chapter 13).

5.2 TERRESTRIAL ECOSYSTEMS

Terrestrial ecosystems can be divided, mainly on the basis of the prevailing vegetation type, in four basic categories: forest, grassland, desert and tundra. The forest is divided into the sub-categories temperate, tropical and northern forest. There are also other categories of lands covered by high natural vegetation that are characterized by a smaller expansion, such as scrubs of temperate marine climates with winter rainfalls (Mediterranean basin, coasts of California and Chile, regions of Australia). The type distribution of terrestrial ecosystems in the planet's surface is determinatively dependent on the average temperature and average rainfall of each area.

5.2.1 Forest Ecosystem

Since the range of temperatures that allow forest development is exceptionally wide, a series of forest types succeed one another in different geographic latitudes. Forest ecosystems correspond also to a wide range of humidity values, from dry to very humid regions.

Further north of the northern hemisphere's temperate forests, a wide forest zone is formed, called *taiga*. It mainly consists of coniferous trees of the *Pinus, Picea* and *Abies* genera. Diversity is small; usually only one or two tree species dominate, while rarely can one find deciduous trees. In this forest zone that surrounds the planet, approximately one-fourth of the biomass and perhaps half of the biosphere's carbon is contained; therefore it constitutes a large store for the surplus of CO_2 that is produced by human activities. The influence of this zone on the climate and the air renewal of the northern temperate regions is also significant. Some characteristic species of big mammals (bears, wolves, deer, reindeer) live in the taiga.

Temperate forests generally grow in regions where the minimum temperature does not usually fall below –12°C and yearly rainfall is around 750–2000 mm. Species diversity is high. Trees are mainly deciduous but there are also coniferous species. Consumer organisms include a great variety of animals. The predominance of deciduous trees exerts a great influence on the character formation of the temperate forest. It allows the development of rich secondary vegetation under the trees, which flourishes in the beginning of spring before the tree leaves cast their shadow on the area, and is a significant producer for the ecosystem's trophic chains. In the temperate forest, due to the falling of leaves, a soil that is rich in nutrients is created that in its turn supports complex decomposer communities as well as those of various small-sized organisms. Thus the soil—that initially is grains coming from rock weathering—ends up being a composite ground ecosystem that contains microorganisms, plants and animals.

The tropical forest, what we know as jungle, develops in the equatorial zone, in regions with high average yearly temperature and rainfall. It includes either ombrophilus vegetation with evergreen, broadleaf trees, where humidity is high for the duration of the year, or vegetation with trees that lose their leaves during the less humid period. The tropical forest, especially the rain forest, is characterised by an exceptionally great diversity of plants and trees. In a few hectares, there may be more plant and insect species than in all Europe's flora and fauna. In many places, more than 50 tree species may be found in one hectare. The tallest trees can reach a height of 60 meters and they often touch each other in their tallest parts thus creating a dome of 30 meters high or more. Very little light reaches the ground through this dome with the result that the secondary vegetation under the trees is sporadic. There may also be a wide variety of epiphytes that grow high on the trees, in the lighted zone, with roots that do not reach the ground. The Amazon forest is a characteristic case.

The ratio of leaf biomass to new wood production is approximately 1:1 in the tropical rain forests, which means that they direct a much smaller part of their production to wood storing compared to other forests. In most tropical forests the soil is thin and poor in nutrients, because the nutrients are washed down to the underground water table by the profuse rainfalls. Less than 50% of the nutrients are found in the soil, versus 90% in the European forests. The effective recycling and symbiosis mechanisms keep the nutrients in continuous circulation inside the forest's biomass. Through the decomposition of leaves, nutrients are released that are very quickly taken in by the shallow roots of the trees. If the forest is cleared, those adaptive mechanisms are destroyed and the use of the land for cultivation or as pasture is minimally profitable.

Scrubs with evergreen sclerophyllous plants, known as maquis, are a characteristic forest vegetation of dry regions, where rain falls mainly in the winter, whereas summer is hot and rainless. Those ecosystems are variable. Frequent fires are an environmental factor to which they are adapted by their nature.

5.2.2 Grassland

In grasslands, trees are absent or they grow sparsely. Rainfall presents intermediate values, between those of woodlands and those of deserts, i.e. 250–600 mm in the temperate zone. Large grasslands appear in the interior of North America, of Europe and Asia, in Central and South Africa, in the southern part of South America, and in Australia. Probably many of Russia's steppes as well as tropical savannahs were once forest zones where the action of fires combined with intensive grazing from herbivores have prevented the return of the forest.

Mammals and birds that can be found in grasslands exhibit great variety (e.g. bison, antelope, wild horse, zebra, rhinoceros, lion, ostrich). Vegetation mainly consists of short-lived plants and a large amount of organic matter is continuously added to the ground. The transformation of forest litter to humus is fast but the decomposition of humus to inorganic materials is much slower. Thus the grasslands' soil may contain 5–10 times more humus than the forest soil. This richness has led to the development of profitable agricultural exploitations. The subsoil part of the plants forms, together with the soil, a protective carpet that prevents soil erosion caused by the wind and runoffs. If the protective carpet is destroyed by ploughing, rapid erosion may follow.

5.2.3 Tundra

Beyond the taiga, the treeless plains of tundra develop in a zone around the Polar Regions. In those areas, where the average yearly temperature is below –5°C, the ground defrosts in the summer only at a depth of less than 1 meter. Atmospheric precipitations are less than 250 mm/year but due to low evaporation the limiting factor usually is not the water but temperature. There is ecological similarity between the tundra and mountain regions with high altitude, above the tree line, the so-called *alpine zones*.

Tundra's productivity is not very low because a large number of species have developed admirable adaptations in order to survive in the cold and take advantage of the short period when the temperature is higher. Primary producers are mainly lichens, mosses and weeds. Reindeer, wolves and polar bears are some of the mammals that live there. A variety of migratory birds nest in the summer and feed on insects and freshwater invertebrates that exhibit a great seasonal growth.

5.2.4 Desert

Deserts may appear in areas with yearly rainfall less than 250 mm or in more humid warm regions with an unequal yearly distribution of rainfall. Most are found between latitudes 30° N and 30° S, although special local conditions can create deserts elsewhere, too. In some extreme cases of deserts, draught can last for many years. The water is usually the limiting factor in the desert; that is why a desert's productivity is more or less a linear function of rainfall. Polar zones are types of deserts where the limiting factor is not the water but temperature; Antarctica is considered as the largest desert in the word.

Plants and animals of the desert have developed many specializations for water economy. Thus thick, waterproof trunks are observed in trees, as well as a reduced surface or a special making of leaves (e.g. special trichome) that helps the reflection of light. The plants may grow far apart but a shallow root system develops in the ground so that absorption of rain can be maximized. Under special conditions of the soil, some plants develop deep roots that reach the underground water table. Animals solve the problem of water deficiency in many ways. Most of them become active at night, while during the day they find shelter in places that offer coverage against high temperatures. Their excretory systems are suitable for water retention, and many animals are able to use the water that is produced during cell metabolism.

5.2.5 The Forest as a Natural Resource

The forest is a complex and rich ecosystem as well as a valuable renewable natural resource. Its direct offerings include:

- Forest biomass, as a fuel that is an energy source of decisive importance, especially for populations of some developing countries
- Timber and a series of other forest products that are still irreplaceable raw materials
- The store of genetic material that can play the role of genetic reservoir for future improvements of agricultural production and of other productive sectors
- The most significant, along with the oceans, shelters of the planet's wildlife
- The use of areas for recreation and enjoyment of nature, which is a vital need of the inhabitants of cities in our times

The indirect offerings of forests are perhaps even more important for man and include:

- Protection of drainage basins against erosion and the creation and preservation of soil

- Regulation and stabilization of the water cycle
- Global climate stabilization through binding and storing of atmospheric CO_2
- Regulation of the local climate through the increase of evapotranspiration and, consequently, of humidity, and reduction of the temperature variations amplitude
- Regulation of nutrients' circulation in the ecosystems
- Regulation of water and atmospheric air quality

Forest destruction has both local and more general consequences. At a local level, the ability of the area to retain underground water is decreased; thus the enrichment of the water table and the spring water supply decrease, whereas the torrent peaks increase. Erosion leads to loss of soil, heavier torrential phenomena and transport of great quantities of debris, while the elimination of nutrients from the ground enriches aquatic ecosystems and can create eutrophic conditions. Flora and fauna variety is significantly decreased and the function of other ecosystems is disturbed. Disturbances are also caused to the local climate and to the water and air quality. At a planetary level, forest losses entail the strengthening of the greenhouse effect and of the climatic change danger, the serious reduction of biodiversity, acceleration of desertification and perturbation of biochemical cycles.

5.2.5.1 Deforestation

The destruction of forest ecosystems, with tropical forests as a spearhead, is one of the major contemporary environmental problems. The phenomenon is more acute today but is not new, given that deforestation began in the Palaeolithic era and has continued throughout human history. Forest clearing for the creation of agricultural land and pastures has played the most significant role. It is estimated that human interventions have reduced the planet's forests to 4.2 billion hectares, i.e. to approximately two-thirds of the acreage they covered 10,000 years ago. The consequences, mainly erosion and floods, were already known in antiquity to the civilizations of China, Mesopotamia and the Mediterranean. In Europe, many forested areas were destroyed during the Middle Ages and the Renaissance under the pressure of the developing population's needs for ship-building material, fuels, lands for grazing or cultivation etc.

Today, the percentage of forest cover has been stabilized in Europe, while deforestation has tended to cease also in most of the developed countries. The global forest problem is mainly centred in tropical forests that continue to be felled or to burn much faster than they are replaced by natural regeneration or artificial reforestation. Every year, approximately 11 million hectares of woodland are destroyed by felling or fires and the forest is restored in only 10% of that surface. The major factor for deforestation is still the effort for the creation of agricultural land, while other significant causes are the demand for firewood for cooking and heating, the creation of pastures for livestock farming, and the exploitation of tradable timber. The change of land use and fires constitute threats for certain forests of countries with a temperate climate, whereas recently, the forests of Northern and Central Europe faced the acid rain threat.

Coping with major global problems, such as the greenhouse threat, erosion, desertification, water shortage and energy deficiency, requires a gigantic effort for upgrading the planet's forests. Programmes are needed for the preservation and improvement of the existing forests, for the extensive reforestation of areas that have suffered deforestation, and for forestation (creation of new forests) in areas where other uses are not necessary. It is estimated that an immediate forest restoration is required, especially in tropic areas, in a stretch of at least 150 million hectares. A basic necessary resource for this endeavour is not only the capital but also work, which only the local farmers can provide on such a scale.

5.2.6 Natural Vegetation in a Mediterranean Environment

5.2.6.1 Forests

Pine forests of low altitudes occur from the coastal areas up to 1200 m; they include several species of pine, cypresses or cedars. These forests develop in Mediterranean climatic conditions and are extremely vulnerable to fires.

In low altitude areas, with more humidity and lower winter temperatures, mixed deciduous forests develop that include mainly various species of oaks.

In stream beds, on river banks and on lake shores as well as in lowland areas with high humidity, like river mouths, hydrophilous clusters of trees develop, such as plane-trees, poplars, willows and tamarisks. The latter often appear along beaches as well.

Mountain forests mostly occur between 500 and 2000 m, in climatic conditions characterized by low winter and temperate summer temperatures, snowfall and frequent rainfall. They are distinguished into coniferous forests that mainly include *Pinus nigra* and species of firs, and into deciduous forests that include beech in higher altitudes and various species of oak or other trees in lower places.

During the past century, forest cover of many European countries has increased; for example, in Greece, it exhibits an ascending course in the last decades, given that the percentage of forests had previously been estimated as 19%. In spite of the limited reliability of the old estimations, the comparison reveals that the destruction of woodland due to fires, over-grazing etc. has been overbalanced in our days either by the natural regeneration and artificial reforestation of the burned areas or, mainly, by the natural forestation of mountainous agricultural lands whose cultivation has been abandoned. However, this tendency will probably be reversed in a future context of climate change. The forests that are burned and are in recession are mainly peri-urban, coastal and insular; they are mostly pine forests at low altitudes. Nevertheless, the quality of many forests is degraded due to the maltreatment (frequent wars, over-grazing, nomadic cultivation, fires, felling) that they have suffered for centuries.

5.2.6.2 Scrubs

The Mediterranean climate is characterized by summer draught, rainfalls mainly in the winter, and strong solar radiation. Vegetation is also characterized by its adaptation to the summer dryness, which may constitute a limiting factor. Mediterranean ecosystems exhibit a wide biodiversity and are divided into the following categories:

- Evergreen sclerophyllous bushes (maquis) that can reach a height of some meters and extend from the coast up to an altitude of 800 meters. They include kermes-oaks, lentisks, holm-oaks, strawberry-trees, carob-trees, wild olive trees, junipers, cedars, laurels, Spanish brooms etc.
- Low bushes with thorny branches and small leaves (phrygana) that grow at the driest regions as well as on degraded soil. They include thyme, Jerusalem sage, rock roses etc. Among the phrygana we find many aromatic plants as well as bulbous plants, like asphodel.

5.2.6.3 Alpine Meadows

In mountain areas above 2000 m, where the average temperature is between 0°C and 10°C and atmospheric precipitations usually exceed 2000 mm, we find the alpine ecosystems that can also occur in lower mountains, if the forest limits recede for various reasons to a lower altitude. They include small bushy and herbaceous plants, some of which have brightly coloured flowers in the summer.

5.2.6.4 Forest Fires and Restoration

Fires are one of the major threats for the forests and every year they cause huge catastrophes to the whole planet. However, it should be taken under consideration that the forest, as an ecosystem, especially in Mediterranean climates, is in a dynamic equilibrium with fire, which constitutes a natural factor that necessitates the corresponding adaptation. Forest ecosystems possess remarkable restoration mechanisms after the fire. Thus, if natural regeneration is not hampered by grazing, a second fire or other human interventions, it can ensure the restoration of the forest vegetation to a large degree. The need for artificial reforestation arises only in cases where natural regeneration does not succeed, for various reasons. Therefore artificial reforestation, which is a difficult and extremely costly activity, plays a rather limited role in forest restoration after fires.

In Greece, the phenomenon of forest fires has exhibited a significant increase during the last 40 years and constitutes the major deforestation factor. A great percentage of the disasters (30–40% of the number of fires and 60–70% of the burned areas' surface) is due to arson, which proves that the effort for land use change towards the direction of pastures and building plots is the most important factor for fires. Of course, we should not ignore the periodic draughts and the potential climate change that create favourable conditions for a greater number of fires and more destructive ones.

5.2.7 Soil and Agricultural Activity

5.2.7.1 Erosion and Desertification

By the Neolithic era, deserts had already expanded significantly due to human action. Deforestation, over-grazing and cultivation of friable soil are basic factors. It is possible that in certain eras some small climatic changes assisted this process and accelerated deforestation. However, the prevailing views suggest that human activity was much more decisive than the influence of the climate. The Nile valley case is characteristic. For thousands of years, it had the ability to feed Egypt's inhabitants, because the ecosystem's functions, with the periodic risings of the river's level and the transport of nutrients, had remained undisturbed. This was so because the sources of the Nile were mostly in inaccessible areas that had not received strong human interventions until recently. On the contrary, in the case of Tigris and Euphrates Rivers, degradation is great. Their sources are in the Armenian plateaus, which have suffered the impacts of human activities and especially those of extensive livestock farming for long periods. The gradual destruction of natural vegetation by over-grazing, timber felling and thoughtless cultivation caused extensive erosion. The river waters are overloaded with debris that deluge the valleys and create extensive alluviums at the river mouths. The old empires that developed in the fertile land of Mesopotamia managed to decelerate the degradation for hundreds of years by land reformation works. However, when the invaders shattered the strong and organized states, the irrigation system was destroyed and the area entered a gradual desertification process.

In vulnerable watersheds where the conditions are favourable, traditional agricultural intervention can destroy a region's productivity through erosion. China's examples are characteristic, where the large population and extensive exploitation caused degradation and disaster due to erosion of one-third of the country's surface centuries ago. The cases of Mediterranean countries are similar. The mountain areas have lost the largest part of their soil in the course of centuries. The causes are the same everywhere: thoughtless tree felling, fires, and above all, over-grazing. The rates of the degradation differ, though, and are related, inter alia, with historical factors such as wars, invasions of various raiders, the actions of pirates, the presence of organized states that protected the public works and were setting limits to over-exploitation etc.

The extreme pressure of grazing operates mainly in mountain areas with steep inclines. The number of animals per surface unit plays a decisive role, because when it exceeds certain limits, vulnerable natural vegetation does not have the time to recover from the excessive exploitation and becomes degraded. The kind and composition of phytocoenoses change, and gradually the ground becomes bare, which reduces its resistance to erosion. The footsteps of the animals that continuously tread the ground cause serious erosion along the routes they follow. Starting from those sensitive zones, the erosion process continues unrestrained towards the neighbouring points.

The most catastrophic combination is the one of grazing following a fire. The natural regeneration of vegetation is greatly sought after by herbivorous animals and gets destroyed; as a result, the system's ability to recover is counteracted after the fire. In this case, not only over-grazing but even limited grazing can have catastrophic consequences. The results are the same in the case of two consecutive fires in a period of a few years.

Erosion can act even in lowland areas with small inclines. This horizontal erosion carries away the fine soil particles and amasses them at the lowest point reducing the land's fertility. Ploughing along vertical contours, the large extent of the fields and the abolition of plant hedges create favourable preconditions for the erosion of even almost horizontal surfaces.

Wind erosion is also significant in open areas, like the steppes. It is caused both by over-grazing and the cultivation of marginal grounds and it is profoundly manifested after long periods of draught. The kind of cultivation also plays a significant role, because friable earth needs proper vegetative cover in order to resist the wind or the surface water flow.

The damage caused by the erosion of cultivable lands of the planet is huge. The phenomenon has been studied especially in the United States, where in 1950 it was realized that two-thirds of the cultivable land showed signs of erosion. It is estimated that in the duration of a century, an expanse of land that equals one-fourth of the lands that can be cultivated was degraded by erosion in the whole planet. Therefore the preservation of the productive potential of the ground that is already under cultivation should constitute a basic priority of agricultural policy on a global scale.

5.2.7.2 Transformation to Laterite

In the warm and humid tropical regions, the soil is subjected, besides erosion, to the danger of its transformation to laterite. Laterites occur in all tropical regions and are characterized by a high content of iron and aluminum, a small quantity of silicon and an acid reaction. Many types of rock can be transformed to laterite after they are decomposed under the influence of continuous leaching out by pelting rains in combination with the high temperatures, due to intense solar radiation. Water penetration causes the removal of the more soluble elements, such as sodium, potassium, magnesium and silicon. Thus the concentration of the less soluble elements, especially iron and aluminum, increases. Laterite is usually a few meters deep. It may occur in non-tropical regions also, but in that case it is a product of previous geologic periods.

The causes of laterite formation are connected with the removal of vegetation. Tropical forests can also develop in infertile soil, where the extraordinary bloom of life is due to the quick and almost perfect recycling of nutrient salts, thanks to the high temperature and the abundance of light and water. When the primary tropical forest disappears by fire or clearing, the soil suffers the consequences of intense leaching; as a result, its structure is altered and laterite is gradually formed, which renders the growth of plants almost impossible. The tropical forest degradation process is very fast, because the layer of dead organic matter (leaf coverage and humus), that in northern forests can reach a depth of a few meters, does not exceed a depth of a few centimetres since decomposition takes place almost immediately. Thus it is possible that large expanses of forests could be irrevocably lost in a few years; moreover, those expanses become unsuitable for any agricultural use and even for a potential reforestation effort.

5.2.7.3 Salination of Soil

Besides erosion, the cultivation of soils with friable structure can cause irreparable chemical alteration, which renders them unsuitable for exploitation and obstructs even the recovery of the initial vegetation.

Soil salination is known since antiquity as a great scourge for agriculture. It comes from inadequate drainage of cultivations or from the presence of a high level water table in combination with profound evapotranspiration favoured by draught and high temperature. The result is the accumulation of salt on the top-soil and the formation of salty plaques. In some cases, irrigation, by leaching out the soil, causes an irreversible change of its structure, thus making the problem even more serious. Calcium is removed and substituted by sodium that moves towards the top-soil during the dry period. The formation of sodium carbonate increases the soil's pH to high values, leading to the scattering of clays and decomposition of organic matter. The soil blackens and loses all fertility because the clay-humus complex has been destroyed. The soil is now useless for any productive exploitation.

5.2.7.4 Agricultural Ecosystems

Some extensive agricultural cultivations form ecosystems that retain enough naturalness. Olive groves, vineyards, grain farming etc. are such examples, when chemical inflows, especially pesticides, are not used in great quantities. In such cases, considerably large and varied natural populations of flora and fauna, e.g. birds, live in this anthropogenic environment and use it for food, nesting, breeding etc. Many stretches of land that extensive livestock farming activity uses as pastures are natural ecosystems, as are many forests that are under exploitation. The distinction between natural and anthropogenic environment in the farming area is often vague.

5.2.7.5 Plant Hedges in Agricultural Areas

Agricultural cultivations were traditionally separated from each other by plant hedges, creating an interesting and varied landscape. The presence of trees, bushes and various small plants, of small clusters of forests or scrub among the fields not only creates a more pleasant picture of the rural landscape but also supports significant diversity of flora and fauna species that has proved to be beneficial for agricultural production. Among others, many birds and various other species, which control some populations of insects and parasites that harm cultivations, nest or find shelter. The interruption of cultivation continuity and the parcelling out of vast agricultural expanses can play a positive role in containing agricultural epidemics. Plant hedges play a significant role in the protection of agricultural soil against erosion. Moreover, the presence of natural valleys and streams between cultivations contributes to the biological and hydrological equilibrium of the rural areas.

The loss of plant hedges has significant negative consequences on the ecological equilibrium of agrosystems as well as on the micro-climate of the rural areas. Plant hedges are usually the victims of rural area modernization that promotes modern intensive agriculture, often regardless of the ecological and long-term economic cost. Land consolidation and land improvement schemes, when implemented indiscriminately and without an expert integrated study, destroy the small natural pockets that are adjacent to the cultivations, and abuse the rural landscape from an ecological and aesthetic point of view.

5.3 AQUATIC ENVIRONMENT

5.3.1 Marine Ecosystems

The continuous interaction of water masses with each other and with the sea bottom leads marine organisms to close relationships with the total of the marine environment (trophic, reproductive, nesting dependence); therefore, it is not an exaggeration to consider the whole marine space as a unified ecosystem.

For practical reasons, it is advisable to distinguish at least two great compartments in the marine environment: the sea bottom's ecosystem and the ecosystem of the water body. The presence or absence of light constitutes a decisive criterion for the distinction of partial compartments, because this is what controls photosynthesis. Primary production either by phytoplankton or by phytobenthos constitutes the energy basis of every marine trophic web; it also supplies with food the great depths where complete darkness prevails. Another important criterion for the compartmentalization of the marine ecosystem is the distance from the coast. Land's influence on the marine environment is often important; it distinguishes the coastal marine ecosystem from the ecosystem of the open sea.

The ecosystems of the sea bottom can be divided into (a) the ecosystems of the continental shelf, that begin from the coastline and reach the beginning of the continental rise, in a depth of about 150–180 meters and (b) the ecosystems of the great depths that begin from the continental rise and continue down to the abyss plains and hadal trenches (maximum depth almost 11,000 meters). Correspondingly, the ecosystems of the water body can be divided into (a) the ecosystems of neritic (coastal) waters, found above the continental shelf and (b) the pelagic ecosystems, found far away from the coasts, beyond the continental shelf.

5.3.1.1 Ocean

The ocean is the oldest, largest and more stable ecosystem of the planet. Its function is determinately influenced by natural factors, such as surge, tide, currents, salinity, temperature, pressure and intensity of radiation. Trophic chains begin with the smaller known producer organisms and end with the biggest animals of the planet. In great depths, near geothermal springs, isolated ecosystems develop with unique characteristics in which primary production is not based on photosynthetic organisms but on chemosynthetic bacteria.

5.3.1.2 Estuary

Coastal marine systems are generally much more productive than those of the open sea, because they are much less limited by light and nutrients. The most characteristic examples are coral reefs, coastal wetlands and estuaries.

An estuary is a semi-closed marine system, as a river mouth in an oblong bay, where salinity assumes intermediate values and the action of the tide plays an important role in the regulation of natural parameters and in the energy supply to the system. Primary production is high and is due to three categories of organisms: phytoplankton, benthic microflora and aquatic plants. Thus, favourable conditions for the development of abundant consumer organisms both in the estuary and the neighbouring ecosystems are created.

5.3.2 Freshwater Ecosystems

5.3.2.1 Water Course

Lotic ecosystems include big and small rivers, creeks and intermittent streams. From an ecological point of view, they belong to two categories: sections of rivers that erode the banks and the bottom, and those that deposit sediments and therefore have a muddy bottom. A river can generally be divided in three parts: the sections of tributaries near the springs, where the water is clear and the concentration of nutrients is low, the intermediate sections where biodiversity and photosynthesis are greater, and the section near the mouth where flow is slow and transparency is low, at least in big rivers. The lotic ecosystem usually has a lot of particularities and its various uses by man, such as water supply, waste disposal, energy production, piscatorial production and flood control can have destructive consequences and need careful multidisciplinary study.

5.3.2.2 Lake

Lakes are collections of stagnant freshwater or, more rarely, salt water inside the boundaries of continents. A lake is directly dependent on its drainage basin; specifically it is dependent on the size, geology, erodibility, topography, land uses, vegetation cover etc. of its watershed. The chemical status of a lake depends on the climatic, hydrological and geological conditions as well as the pollution of the drainage basin, in conjunction with the biological condition of the water ecosystem, which is connected with the concentrations of dissolved oxygen and nutrients, transparency, primary production, presence of higher predators etc.

5.3.2.3 Wetland

Wetlands' ecosystems include marshes, relatively shallow lakes, lagoons and other coastal aquatic systems, as well as riparian or hygrophilous forests. Their productivity is generally high, as in the case, for example, of estuaries. The hydrological function is a determinant for all wetland systems.

Biodiversity, especially that of fauna, is high and their general ecological role is important. They constitute significant regulators of the carbon and phosphorus cycles, as well as of the nitrogen cycle, because they significantly contribute to denitrification. They play a significant role in the hydrological cycle because they contribute to the supply of underground water-bearing formations and reduce the impacts of floods. Many studies show that in the long run, wetlands have a greater economic value if they remain in their natural state than if they are drained or turned over to other uses.

5.4 COASTAL ZONES

Coastal zones, which include the continental shelf down to a depth of about 200 m and the maritime plains with an altitude less than 200 m, constitute about 8% of the Earth's surface. The biological wealth of the oceans is mainly concentrated in the coastal areas where very productive ecosystems, such as wetlands, submarine meadows and coral reefs, meet. Eighty percent of the global fishing production comes from the coastal zones, while 60% of the world's population lives in places that are less than 60 km distance from the coast.

5.4.1 Characteristics of Coastal Ecosystems

In a very general ecological view, the coast is included in the wider zone of the lithosphere's contact with the hydrosphere and is a product of their everlasting antagonistic action that is formed through their interaction in the course of time. The coastal zone is an expression of the dynamic equilibrium between sediments' supply by the lithosphere and their arrangement (removal or accumulation) by the hydrosphere by the action of waves and currents, factors that are to a great extent dependent on the atmosphere. Also, the biosphere's influence (biogenic coasts, bioerosion etc.) in the formation of the coastal zone is not negligible. Man's action in the coastal space by various works is often extremely important; those works, in combination with the natural characteristics of every area, create a new, anthropogenic, dynamic equilibrium.

During geological eras, the fall or rise of the sea surface (e.g. during glacial or interglacial periods) as well as the perpendicular movements of Earth's solid crust (due to tectonism or other geological causes) had as a result the emersion or submersion of the coasts. The form of coasts changes continuously as the liquid element adapts to the shape imposed to it by the continuous continental drift. This drift (explained by the theory of lithospheric plates), starting from a unified continent (Pangaea) and a unified sea (Panthalassa) led to the current form of the global coastline. From a geological point of view, today's coasts are recent formations because they were formed in the time space between the Palaeogene and Neogene periods, with outlines similar to the contemporary ones. They were finally shaped in the Quaternary, circa 2 million years ago.

The coast is a characteristic example of an ecotone, i.e. of a transitional phase between two different ecosystems. It constitutes an intermediate zone between land and sea, with an extreme linear geometry, given that it has a very great length and small width, which makes it very vulnerable to external perturbations.

The ecological characteristics of the coast come from the overlapping of the ecosystems it divides: for example, it has an intermediate temperature, humidity, or salinity, while it includes biological species of both marine and terrestrial environments as well as characteristic species of coastal habitats. As is the case in other ecotones, species diversity and density of biological populations are usually greater than those of the neighbouring areas. Productivity is also generally significant, because the main limiting factors, which are nutrients in the sea and humidity in the land, usually present a satisfactory availability in coastal ecosystems. Thus the coastal environment bustles with life, given that trophic conditions are satisfactory for the organisms.

An integrated knowledge of the coastal environment, however, is not at a satisfactory level and there still is a significant lack of scientific information. This is partly due to the fragmentation of research about partial objects, e.g. terrestrial environment, marine environment, botany, zoology, geology etc. There is a great need for interdisciplinary projects so that the function of coastal ecosystems and their conservation needs as well as the possibilities of their utilization can be studied.

5.4.1.1 Coastal Wetlands

Some coastal ecosystems, such as the areas of coral reefs, of brackish marshes, lagoons, river Deltas and tide zones have a very high productivity, comparable to that of tropical forests. These ecosystems belong to the category of wetlands and greatly contribute to the ecological equilibrium of the neighbouring terrestrial and marine ecosystems. Coastal wetlands are very important for feeding, resting, nesting and breeding of a large number of aquatic, seashore, and marine, migratory or non-migratory, birds, as well as of many birds of prey, and also for the preservation of many biological species that are endangered. They also play a determinative role in the development of marine organisms' early stages and especially the alevin of many fish that are of great significance for fishing.

The role of these spaces is also significant in the decomposition of dead organic matter, in the completion of processes in the cycles of carbon, nitrogen, and phosphorus, and in the cleansing of incoming waters from nutrient salts, organic substances, heavy metals etc. Thus they significantly contribute to the decrease of marine pollution since they have the ability to break down and retain, through physical or biochemical processes, the pollution loads contained in agricultural runoffs and in urban and industrial waste. Of course, their assimilative capacity is not unlimited and they face the risk of serious degradation if the pollution exceeds certain concentrations, according to the kind of pollutant. This sensitivity increases when the depth is small and the water renewal is slow.

5.4.2 Coast Degradation Factors and Protection Needs

Coastal ecosystems, due to their special characteristics, have increased sensitivity and require rational management. They constitute a significant ecological and economic capital with great possibilities if properly utilized, which, however, can quickly suffer degradation because of thoughtless interventions. Since the first developmental stages of societies, man was closely connected with the sea. Most settlements and the largest concentration of people developed in the coastal areas. First, the sea ensured a percentage of man's diet, while with the development of trade it became the main route for the traffic of products. Moreover, the climate in coastal areas is milder and friendlier for man's subsistence. Man's traditional relationship with the sea was to a great extent harmonious without causing serious negative impacts to the environmental equilibrium. This relationship, though, was gradually disturbed because of the subsistence needs of the increasing human population as well as the multiplying consumer needs. The development of industrial production and the rocketing increase of technological abilities seriously undermined man's relationship with the sea, mainly because of the uncontrolled dumping of all kinds of waste into the sea from the coastal urban and industrial areas as well as from sea transport.

The development of technology improved man's living standard and his adaptability to various environments; nevertheless a large percentage of the world's population still lives in the coastal areas. Most big cities, where the greatest part of economic and other activities is centred, are near the coast. Significant percentages of raw materials and energy are obtained from the coastal region: oil and gas, wind, wave and tidal energy, seafood from fishing and aquaculture, sand and gravel from marine aggregates, salt from sea saltworks, freshwater from desalination, etc. The development of new technological applications creates prospects for even more intensive exploitation of the coastal and marine space. At the same time, the lack of proper management, the depletion of natural resources and the accumulation of waste in the environment by production and consumption activities create serious degradation problems in the coastal region; these problems often become a hindrance for developmental activities and a threat for people's health and quality of life.

Environmental threats are, as a rule, due to high concentrations of human population and activities as well as to the fact that the coast is usually the receptor of many environmental disturbances that have their source either at the inland or at sea. The environmental pressures that come from intensive human activities cause a rapid degradation of both the terrestrial and the marine part of coastal ecosystems. Basic factors of coasts' degradation are agricultural, industrial, residential and tourist activities, overfishing, as well as the technical infrastructure works that disturb coastal ecosystems to a great extent. Pollution and high pressures on the coastal and marine environment in particular, degrade the natural ecosystems with further multiple negative impacts on the trophic resources, health, the survival of biological species, even on recreation and tourism; moreover, ignorance about environmental parameters when designing coastal works or river dams may cause, inter alia, coast erosion. The coastal zone receives pollution loads through direct waste disposal as well as indirectly by the rivers and the atmosphere. It is estimated that more than 75% of marine pollution comes from terrestrial sources. The coast's condition is also influenced by changes in salinity or the transport of debris and nutrients that are caused by river dams, deforestation and other changes of land use. One more threat is the probable rise of sea level as a result of the amplification of the greenhouse effect.

5.4.3 The Mediterranean Coastal Environment

Mediterranean coasts, in comparison to the Atlantic or the Pacific Ocean coasts, present two main differences that are especially important for their management. The first is the generally oligotrophic character of the sea that washes them, and the second is the lack of noteworthy tide. The oligotrophic character of the Mediterranean allows its characterization as "semi-desert" and is summarized in three elements:

- Small concentrations of nutrient salts (N, P, Si)
- Seasonal and limited development of phytoplankton and phytobenthic populations
- Rare creation of anoxic or hypoxic conditions in the bottom and very rarely in the water body

The oligotrophic character of the Mediterranean determines to a great extent the effectiveness of man's interventions in coastal problems, such as urban wastewater management, the use of fertilizers in agriculture or even the construction of river dams. The lack of significant tide intensifies even more the oligotrophic character of the Mediterranean Sea, because it deprives the marine environment of the everyday stirring of the bottom caused by the tidal current. Moreover, it simplifies the technical parameters of coastal works (harbours etc.), given that fluctuation around the mean level is usually insignificant.

The best criterion that has been proposed for the ecological demarcation of the terrestrial Mediterranean space is the geographical spread of the olive tree (*Olea europea*). Mediterranean climate constitutes an intermediate type between the sub-tropical and the temperate zone and is characterized by mild temperatures, low rainfall, summer draught and a great atmospheric clarity. The influence of the climate marks off the boundaries of mature ecosystems, which, however, are very few in the Mediterranean coastal space due to man's long-term influence.

In the border of the terrestrial and marine environment, a mosaic of ecosystems that present a special ecological and often economic interest appears. Mediterranean coastal ecosystems are mainly represented by wetlands, sand dunes, sandy or rocky coasts, sea cliffs, beach forests and scrubs as well as biocoenoses of an insular character. They host variable and significant biological communities due to the special micro-climatic, topographic and ground conditions.

Harsh environmental conditions, such as strong winds, dryness, high salinity and the absence of soil, usually prevail on rocky coasts. Vegetation includes herbaceous plants and small bushes, while several species of seabirds nest in the rock holes, as does the Mediterranean seal (*Monachus monachus*). Rocky coasts may create notable ecological habitat types. Isolated precipitous beaches lend themselves to seabirds nesting or they may host other kinds of bird fauna. Rare species of plants grow also on coastal rocks. The coasts of small islands have special significance where ecosystems are of particular interest and the ecological impact of the inland as well as the human impact are relatively small. In many of the thousands of small islands or uninhabited rocky islets there are rare endemic plants and local endemic reptile species and subspecies (e.g. lizards).

Sandy beaches are sediments' disposal regions and the environmental conditions are characterized by poor soil, high salinity and dryness. Biodiversity in general is not great but the sandy beaches are an ecological habitat for some rare species, like sea turtles (*Caretta caretta*, etc.). Sand dunes are rare in the Mediterranean and they are often remnants of older climatic changes. They are usually connected with torrent and river mouths or come between coastal wetlands and the sea. When they are in a satisfactory natural condition, they are covered by herbaceous halophytic vegetation that may include endemic or threatened flora species. This vegetation is extremely vulnerable and becomes degraded very easily by human activities. In some cases, as for example in the Marathon area, sand dunes host coastal forests that consist of *Pinus pinea* or a mixture of *Pinus pinea* and *Pinus halepensis*. Such forests were common in the past but today they are rare.

Mediterranean coastal wetlands are river-mouths, estuaries, lagoons, closed bays, brackish or freshwater marshes and saline fields, humid grasslands, wallows etc. exhibiting a significant morphological, geochemical and sedimentary diversity. They are characterized by the presence of water or by high humidity. They may temporarily dry up during the summer period, whereas near the end of the winter period they exhibit their greatest development. They are usually characterized by a significant salinity change in connection with the altitude and the distance from the shore.

The Mediterranean coastal wetlands constitute ecological unities because the biocoenoses are not characterized by a mixture of freshwater and sea water species, but mainly by communities with many species adapted to this specific environment and endowed with great tolerance to salinity (euryhaline species). In contrast with the ocean coasts, the tide is negligible and does not constitute a significant factor for the Mediterranean coastal ecosystems. This fact, in combination with the light rainfall, hinders the dilution of the beach lagoons; as a result, ecological habitats with subsaline or often supersaline waters abound in Mediterranean coasts. Vegetation as a rule is very rich, but most of the plant species are common. There is a great variety of fauna, especially birds, that includes many rare species. The wetlands' biomass and biological productivity are noteworthy and their economic significance is considerable. Coastal ecosystems ensure 10–30% of the total Mediterranean piscatorial production and they host many aquacultures. Their sensitive location exposes the Mediterranean coastal wetlands to increased degradation and destruction risks due to the development of tourist facilities, urban and industrial activities, harbour works, airports and beach roads as well as increased pressures for drainage and other land improvement works.

Coastal forests and scrubs, as Mediterranean forest ecosystems in general, today cover only a small percentage of their initial geographic extent. They are mostly ecosystems that have departed from their natural mature (climax) state due to human intervention over the course of many centuries. The main factor of change has always been fire, which has significantly favoured the extension of coniferous forests as well as of scrubs. The pine (*Pinus halepensis* and *Pinus brutia*), kermes-oaks (*Quercus coccifera*), lentisks (*Pistacia lentiscus*), carob-trees (*Ceratonia siliqua*) etc. are characteristic species of those ecosystems. Cases of advanced degradation are frequent, and their characteristic is sparse bushes, phrygana and large bare rocky stretches of land, which generally constitute evidence of repeated fires and over-grazing.

Mediterranean coastal systems are especially sensitive to the changes caused by human activity; as a result, today's Mediterranean coastal natural environment has become mainly anthropogenic. The great recess of deciduous or sclerophyllous oak (*Quercus* sp.) forests during the last 10,000 years and the expansion of coniferous forests (*Pinus* sp.) as well as scrubs that were favoured by human action are characteristic examples. Fires, grazing and the expansion of cultivations have been major factors.

The current phyto- and zoocoenoses of the terrestrial Mediterranean space are the result of a lasting action of geological, paleoclimatic and anthropogenic factors. The presence of a large number of endemic species in certain areas is associated with the existence of many isolated mountains and islands and is to a great extent due to the significant ecological perturbations that were caused by the glaciers and other climatic changes. It is estimated that in the Mediterranean area there are approximately 12,000 species of endemic anthophytes.

5.5 BIODIVERSITY AND LANDSCAPE

The concept of biodiversity includes the genetic variety among individuals of the same species, the variety among species and the variety among ecosystems (ecological niches and landscapes). The biological species variety in ecosystems and indeed in the biosphere has been one of the main objects of the efforts for environmental protection on the part of international organizations. At the International Rio de Janeiro Convention for the Environment and Development in 1992, biodiversity was one of the most important discussion issues, and the convention concluded with the signing of an international agreement for its protection.

The conservation of biodiversity is very important, mainly because the extinction of one flora or fauna species constitutes irreparable environmental loss, since neither any natural mechanism nor technology can re-create it. Recent estimations give great emphasis to the economic value of biodiversity as well. Many wild flora and fauna species could be useful in the preparation of new medicines, the improvement of agricultural and livestock farming production, the utilization of new biotechnology achievements etc.

The estimated number of biosphere species fluctuates between 5 and 100 million, with a most probable value around 8.7 million, while over 1.2 million species have been described by science and already catalogued in central databases. The current species extinction rate is somewhere between 50 and 1000 times greater than the natural rate. In 1996, 25% of mammals were considered endangered, 11% of birds, more than 25% of freshwater fish and a significantly higher percentage of invertebrate animals.

The most significant factors for the global biodiversity decrease are the following:

- The destruction or fragmentation of ecological habitats (feeding and breeding sites etc.)
- Over-exploitation (fishing, timber felling, trade etc.)
- Introduction of alien species into ecosystems

- Water, air, soil etc. pollution and toxic pollution of the trophic chains
- The climatic change prospect

5.5.1 Flora and Fauna

The total of plant species of a terrestrial or aquatic ecosystem is called *flora*. The term *vegetation* refers to the total of plants that cover the surface of an area. The vegetation of forests, meadows and marshes are characteristic cases. The processes, causes and outcomes of natural vegetation degradation are characteristically described in the case of forests.

The degradation of vegetation, and especially forest clearing, entails, among other factors, the decrease of flora variety. During the past centuries, but especially since 1800, a great number of plant species have disappeared from the biosphere as a consequence of man's actions. The destruction of tropical forests and of other significant biotopes, over-grazing, over-collection of medicinal and aromatic plants and the introduction of alien species into sensitive ecosystems are the major causes for the extinction of plant species.

The total of animal species of a terrestrial or aquatic ecosystem is called *fauna*. A range of factors has led man—since the Palaeolithic era up until the 20th century—to cause the gradual decrease of the populations and the extinction of more and more species of wild fauna. Vegetation offers food and shelter to most of the terrestrial animal species; thus, the destruction of vegetation or, more generally, of the biotopes, constitutes, as a rule, the most significant degradation factor of the fauna. Actually, extensive deforestation, which continues today in the tropical regions, has caused the extinction of many mammals, birds etc. in Europe, North and South America, Australia, Polynesia and other regions. Great disasters have also been caused by the drainage of wetlands, and generally the changes of land use in favour of agriculture, livestock farming and residential development.

Over-hunting and over-fishing are significant factors for fauna degradation, which has driven many species of big mammals, birds and marine organisms to extinction. The change of competition conditions, with the introduction by man of alien species, that is, which are coming from other parts of the planet, into virgin ecosystems, also has had adverse consequences. For many sea mammals, big terrestrial mammals and fish, the decrease began long before the industrial era. Man's expansion 40,000 years ago, during the Pleistocene period, caused the extinction of a significant number of animal species because of over-hunting and deforestation. It has been ascertained that the extinction of animals closely followed man's progressive colonization of various continents and islands. It is estimated that mass species extinctions took place about 40,000 years ago in Africa and Southeast Asia, 20,000–13,000 years ago in Europe, 13,000–2000 years ago in North and South America, 800 years ago in Madagascar and 700 years ago in New Zealand. The extinction rates were accelerated since the Roman Era in the Mediterranean area and since the Middle Ages in all of Europe. During the last three centuries, fauna destruction was greatly extended in the colonies founded by the Europeans on the whole planet. More than 160 bird species, 110 mammal species, and an unknown number of fish, reptiles etc. were extinct after AD 1700. The 19th century especially was characterised by a real carnage of the wild fauna in North and South America, Africa and Oceania.

5.5.2 Biotopes

Biotopes are geographical entities where biological species live. Significant biotopes are distinguished by the presence of threatened flora and fauna species in them. Certain species may live permanently in the biotope, and others use it periodically or occasionally for food, nesting, breeding, resting etc. The species characterized as threatened are the ones that:

- Are endangered—that is, they may disappear in the immediate future
- Are vulnerable—that is, they may become endangered if the environmental conditions become aggravated
- Are rare—that is, may become vulnerable because they have small populations

The European Union biotopes, where the most significant of the threatened flora and fauna species live, have been included, based on certain criteria, in a relevant list compiled by the European research programme CORINE (Coordination of Information on the Environment) in the 1980s. A more recent European programme resulted, based on the most important ecological habitats, in a second list of European Union protection areas called Natura 2000.

5.5.3 Pressures on the Natural Landscape

The natural environment is being degraded by increasing anthropogenic pressures. The populations and ecological habitats of the threatened species as well as the landscape features are mostly affected by the following factors, although the extent and the intensity of the dangers these factors pose depend on particular ecosystem conditions:

- Intensive agriculture with the chemical inflows and the exploitation of terrestrial and aquatic resources (pesticides, over-consumption of fertilizers and water, drainages, groundwater pumping, river diversions etc.)
- Pollution from toxic or nontoxic liquid and gaseous waste
- Dumping or disposal of solid waste

- Forest fires
- Over-grazing and resulting soil erosion
- Overhunting and overfishing
- Excessive wood cutting and deforestation
- Surface extraction
- Excessive road-cutting (mainly mountain or coastal roads)
- Vacation houses or tourist developments
- Construction of technical works with insufficient environmental control

Coastal landscapes are particularly vulnerable to human activities and their degradation is, as a rule, non-reversible and for this reason especially serious. In spite of public perception, sea pollution is not generally a serious problem, given that it is only localized in a few coastal areas. The disposal of liquid urban and industrial waste, the accumulation of rubbish, pollution by petroleum products, agricultural runoff, burdening of the atmosphere by transports and coastal industrial activities usually create local phenomena of marine pollution. On the other hand, forest fires, erosion, drainage of wetlands, all kinds of thoughtless excavations, the haphazard expansion of vacation houses and tourist activities, loud noise sources, free camping, uncontrolled use of recreation vessels, over-fishing etc. are factors that exert intense pressures on the coastal environment. Many coastal landscapes and seascapes are immediately threatened by technical works, such as the construction of coastal roads, fishing shelters, marinas etc., but they may also be threatened by the use of campers, power boats, underwater equipment etc. Thus the rare species of flora and fauna that live in certain coasts are significantly receding or are in immediate danger of extinction.

5.5.4 Greek Landscape: A Characteristic Example

5.5.4.1 Landscape Attributes and Alterations

Greece is a typical Mediterranean country with significant forest cover; according to recent data, the forest covers 25% of its area. Twenty-four percent is scrub, 11% is phrygana and grasslands stretches, 1% is aquatic surfaces and marshes, 2% is urban and industrial areas and agricultural land is about 35%. According to these data, the natural landscape corresponds to more than 60% of the country's terrestrial extent. Actually, though, the natural space is quite larger, because many agricultural stretches can be included in it, as for instance olive groves, which are akin to the natural ecosystems. The extended marine space is also mostly natural. Therefore, Greece possesses a natural landscape that constitutes a very high percentage of the national space; it also has an extraordinary biodiversity and it therefore holds first place among the European Union member-states, in spite of the country's small size (Figure 5.1). The degree

Figure 5.1 Important nature conservation areas in Greece: SONB, CORINE biotopes, Natura 2000, and other biotopes. Note that there is a lot of overlap among them. (From Hadjibiros, K. (2007). *Ecology. Ecosystems and Environmental Protection,* 3rd edition. Symmetria, Athens (in Greek). With permission.)

of ecological variety is mainly associated with the following factors that characterize the country:

- It is situated in a special geographic position, at the crossroads of three continents that constitute different biogeographical regions; also, it encompasses significant migration routes of various migratory birds.
- It presents a wide range of climatic types, from hypo-tropic to continental central-European climate. Moreover, the partial coverage of its surface by glaciers during the last glacial period created additional biological diversity.
- It has a complex geological history, and as a result, soils from different rocks are created. There is also a great deal of geomorphological fragmentation that leads to a multitude of islands, mountains, valleys, wetlands and lengthy coastal regions.
- The long and turbulent history of the country, with the occasional abandonment of certain areas by the human population for long periods of time, the extensive destructions caused by wars, the isolation of areas for military purposes, the preservation of undeveloped areas with difficult access etc. has also contributed to the environmental variety.

Table 5.1 Greek Coasts Included in Protection Areas According to European Directives, International Conventions, or Scientific Programmes

Protection Area Category	Total Number of Protection Areas	Number of Coastal Areas	Length of Coasts (km)	Percentage of Coasts' Length (on a total of 16,200 km)
Wetlands of Ramsar Convention	11	8	600	4%
Areas of Bird Directive 79/409 EEC	149	84	2500	15%
CORINE programme biotopes	430	200	5300	33%
Natura 2000 programme biotopes	296	136	4500	28%
Sites of Outstanding Natural Beauty	449	196	2200	14%
Sum total of protection areas	1335	624	8000	49%

Note: There is a great deal of overlap among areas.

Source: Hadjibiros, K. (2007). *Ecology. Ecosystems and Environmental Protection,* 3rd edition. Symmetria, Athens (in Greek). With permission.

The uneven—temporally and spatially—rainfall distribution, in combination with geomorphology, has led to the formation of many and varied drainage basins. There are many small rivers, lakes and swamps, as well as artificial reservoirs. In many of these freshwater ecosystems there is rich riparian and aquatic vegetation and the animal world, especially fish and birds, presents great biodiversity with many rare species. Of special ecological interest are the mountain lakes in alpine regions, as well as river or torrent ravines that have been formed by the erosion of steep calcareous slopes. Marine ecosystems include open seas and closed bays, with a large variety of fish and invertebrates.

The condition of biotopes, flora and fauna species remains very good compared with the rest of the European countries. However, the degradation tendencies are strong and the percentage of threatened biotopes is constantly and rapidly increasing. The fact that many ecosystems are mostly characterized by a small scale, which renders them vulnerable to pressures, contributes to the above. Moreover, a significant part of the flora and fauna species are endemic, that is, they occur only in Greece and not in the rest of the biosphere; their populations are not large and the need for their protection is imperative. The local endemic species, that is, those whose biotope is confined to a unique area of the country, have an even greater need for protection (Table 5.1). A significant number of rare or endemic flora and fauna species and subspecies have engaged the interest of North European researchers, who have been studying them since the 19th century.

5.5.4.2 Degradation of the Coastal Zones

The largest part of the coastline still retains a good environmental quality, since in the past, disturbances were usually small and of short duration.

However, the intense developmental trends of the last decades have created increased pressures on the coastal environment and have initiated a rapid degradation process in many cases.

In countries where the inland consists of massifs, deserts or other tracts unsuitable for intensive development and when there are no other inhibiting factors, like piracy, the coasts play a dominant role in economic and social development. In Greece, this spatial phenomenon has appeared to a great extent since the 19th century and has intensified especially after 1950; as a result, the coasts have developed into its most important economic resource.

It is estimated that more than 70% of the Greek population, 80% of industry, 90% of tourism, a large part of high productivity agricultural land and almost all of fishing activities and aquacultures are gathered in the narrow coastal zone. The antagonistic uses and the impacts of all those activities exert significant pressures on the coastal environment; they threaten certain rare biological species, some sensitive biotopes and many landscapes with significant aesthetic value. The qualitative condition of coastal water and the atmosphere, the ecological balance of many coastal ecosystems, and the touristic value of a large part of the coasts are also under threat.

A characteristic activity that decisively degrades both the biotopes and the natural scenery of the coasts is haphazard land use. Land use conflicts in the narrow coastal zones are extremely intense. The major pressures for land use change are connected with residential development, tourist infrastructure and extension of agricultural cultivations, and less often with industrial facilities, aquacultures and livestock farming or recreation activities. Pressures are exerted on natural ecosystems, usually seashore forests, brushes and wetlands or on agricultural land that is not intensively cultivated. A significant percentage of the coastal natural landscape is taken up by vacation settlements, rooms to let, free camping spaces, coastal dust roads etc.

The most effective tool used for the change of use of the coastal forest land is arson. Successive fires, often accompanied by grazing of the burned areas, constitute one of the major mechanisms that change the natural landscape, destroying coastal vegetation. Various drainage works, by which farming or other activities take up natural expanses of large and small coastal wetlands, have a similar result.

5.5.4.3 Sites of Outstanding Beauty

Nature's wealth is manifested with the presence of great biodiversity. The numerous combinations of ecological parameters create many different types of habitats, and consequently an extraordinary variety of flora and fauna, significant biotopes and natural landscapes. The number of places that have great value as biotopes or natural landscapes reaches several thousand and includes forests, wetlands, coasts, islands and rocky islets, alpine regions, screes, rivers, lakes, ravines, springs, caves etc. There is also an exceptional cultural wealth that is closely connected with ecological variety and natural beauty. The historical monuments of various eras, traditional architecture, manners and customs, traditions and generally all cultural characteristics, together with the natural elements, compose the notion of an aesthetically significant landscape. This can be either an "everyday" landscape that includes productive and other human activities or a "monumental" landscape area.

A Site of Outstanding Natural Beauty (SONB) is a place distinguished for its aesthetic value and which remains to a significant extent natural, although it may be built up. Its size is defined by human measures; it generally should not exceed one day's hiking. SONB often include traditional settlements and archaeological or historical sites (Table 5.2). The criteria for the selection and evaluation of SONB are associated with geomorphological and ecological characteristics, such as relief, vegetation and flora, presence of fauna, water, meteorological conditions, panoramic view, and cultural characteristics such as the presence of monuments, historical reference, traditional land uses etc. Possibilities for recreation and education, the presence of pathways, the feeling of getting away or of retreat, generation of emotion, the possibility of understanding natural processes etc. can also constitute selection criteria. A large part of SONB is threatened by degradation (Table 5.3) due to intense anthropogenic pressures (Table 5.4).

Table 5.2 Geographical Distribution of SONB in Greece

Peloponnese and Ionian Islands	105
Sterea Hellas	54
Thessaly and Epirus	66
Macedonia	53
Thrace and Aegean Islands	89
Crete	82
Total	449

Source: Hadjibiros, K. (2007). Ecology. Ecosystems and Environmental Protection, 3rd edition. Symmetria, Athens (in Greek). With permission.

Table 5.3 SONB Distribution in Relation to the Conservation Status

Excellent or good status and rapid degradation	4%
Excellent or good status and slow degradation	28%
Excellent or good status and upgrading	5%
Excellent or good status and stability	56%
Bad status and upgrading	0%

Source: Hadjibiros, K. (2007). Ecology. Ecosystems and Environmental Protection, 3rd edition. Symmetria, Athens (in Greek). With permission.

Table 5.4 Recorded Reaction of the Local Society
              to Strict Protection Measures of an SONB

Positive	46%
Indifferent	39%
Rather negative	13%
Strongly negative	2%

Source: Hadjibiros, K. (2007). *Ecology. Ecosystems and Environmental Protection,* 3rd edition. Symmetria, Athens (in Greek). With permission.

Chapter 6

The Environment in the City

DOI: 10.1201/b14609-6

6.1 THE URBAN ENVIRONMENT

The problems of the urban environment are primarily associated with the rapid increase in the human population over recent decades. One of the most important indirect effects of this increase has been much more rapid, unplanned and careless increase in the populations of many large cities, especially in the developing countries. In 1970, 40% of the Earth's population lived in cities; by 1999 this figure had increased to 48%, and it is expected to rise to 60% by 2030. It is estimated that during the period between 2000 and 2030 the world population will increase by as many as 2 billion people to the urban population exclusively.

It is obvious that urbanization, especially in developing countries, exacerbates the problems of the urban environment as shown in Figure 6.1. The large concentrations of populations in urban centres along with other factors such as the growth of the urban consumerist lifestyle are serious contributors to degradation of the environment, which includes the following:

- The urban environment has become less natural, due to the decrease in vegetation, fauna and the naturalness of landscape.
- The periurban environment has become less natural, due to the development of transport infrastructure and other activities that serve the city, as well as excessive logging in the developing world cities.
- Solid, liquid and gaseous waste, along with waste energy, are being discarded into the broader urban environment.
- More natural resources taken from a much broader environment are being consumed in order to serve the city.
- Adverse environmental conditions specific to indoor areas are being created.

A range of environmental problems are created in cities over a certain minimum size. The concentration of population characteristic of urban centres is a serious contributor to pollution and other ecological perturbations.

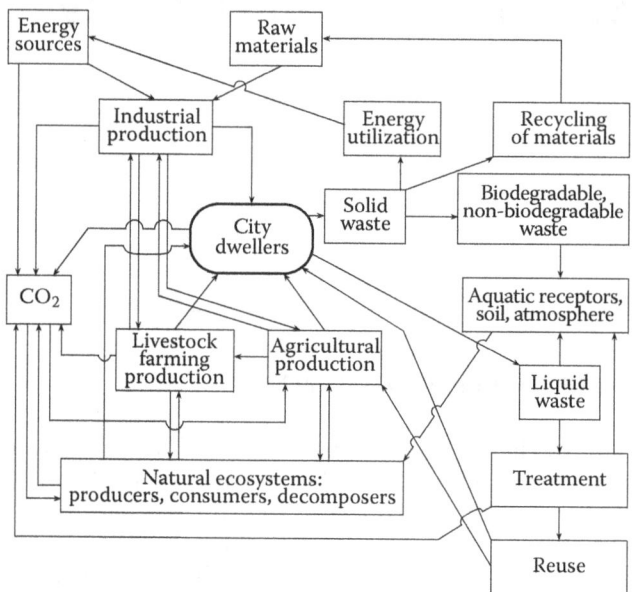

Figure 6.1 Matter and energy flow in the urban system. (Modified from Hadjibiros, K. (2007). *Ecology. Ecosystems and Environmental Protection,* 3rd edition. Symmetria, Athens (in Greek).)

The quality of the urban environment is to a large extent dependent on air pollution (Chapter 9), pollution of surface and underground water (Chapter 8) and the soil, pollution by solid waste (Chapter 6, Section 6.6), and degradation of the natural environment in the urban and periurban space (Chapter 6, Section 6.8), as well as the intensity of other disturbances occurring mainly in cities, such as noise (Chapter 6, Section 6.5), poor quality of air in indoor spaces (Chapter 6, Section 6.7), etc. In addition, many factors that contribute to undermine the quality of life in cities, such as traffic congestion, insufficient green, open and public spaces, aesthetic degradation, risks from accidents or natural disasters, crime rates etc. are directly or indirectly linked to the situation in the urban and periurban environment. Addressing these problems requires adequate environmental management (Chapter 11) as well as use of the appropriate technology (Chapter 10), adapted to the special requirements of more or less developed countries.

6.2 THE CITY AND URBAN ACTIVITIES

The study of the urban environment is part of the scientific field of urban planning, whose main subject is the city. A related concept, but broader than that of the city, is the concept of the settlement. A settlement is a group of dwellings at a relatively short distance from each other, so that certain spatial and functional relationships are created among them. The settlement enters into being from the moment that the number of households making it up is sufficient to create a social group larger than the family. Every city is in the broad sense a settlement, but not every settlement is a city.

In order to be classified as a city, a settlement must have a minimum capacity and certain other characteristics. This capacity must be quantitative as well as qualitative. According to statistical criteria, in order for a settlement to be considered a city it must exceed a certain number of inhabitants. As early as 1887 the International Statistical Institute wanted to classify as a city every settlement with a population of more than 2000. This, however, proved to be impracticable, because each country sets its own threshold in keeping with the special conditions characterizing it, in the way that suits it best. In some countries, settlements with at least 10,000 inhabitants are considered cities, settlements with 2000–10,000 inhabitants are classified as towns and those with fewer than 2000 inhabitants are classified as villages. In special cases, however, a city may be smaller, e.g. the capital of a prefecture.

Number of inhabitants is not enough to distinguish a city from other settlements. There are cases where although the number of inhabitants of a settlement is low, other characteristics (e.g. high density or highly developed commercial, administrative or cultural functions) raise it to the status of a city. By contrast, a numerically large settlement serving only one specific function (e.g. military base or large research centre) cannot in general be described as a city, since it lacks other basic functions. The same is true of certain large rural settlements, especially in developing countries which, despite the number of their inhabitants, are still large villages because they lack characteristic urban activities.

According to modern functional urban planning, the city must serve the four basic needs of its inhabitants: housing, jobs, recreation and circulation. The small settlement which is not yet a city usually meets the first need for housing and only meets the other needs in a rudimentary way. A large part of the work of the inhabitants of a rural settlement is usually performed outside the settlement, i.e. in the fields, whereas other tertiary sector activities find no place in the small settlement and become concentrated in the neighbouring city. In the small settlement even the surfaces available for recreation and circulation are minimal. In contrast, in the city so many services and opportunities for employment are provided that the great majority of the inhabitants are able to live, work, play and communicate without going beyond the city's boundaries.

Another important point that distinguishes the city from smaller settlements is the intensely artificial environment that is created in the city. The more a city grows, the more the space within it becomes alienated from the elements of the natural space surrounding it. Some uban planners argued in the 1930s that we talk about a city, when man prevails over nature and succeeds in freeing himself from it. But this "freeing" of man from nature does not have only favourable consequences. In large cities, along with environmental degradation, certain characteristic physical, mental or intellectual disorders such as alienation and anxiety have been observed, which may be linked to the strongly artificial environment, alienated from nature. Furthermore, the haphazard way many cities and settlements are built often seriously damages the natural environment surrounding them. One can encounter characteristic examples all over the industrial or developing world. This is why one can no longer blindly accept the above argument, which is also deficient since it fails to specify what is meant by "prevailing" over nature or "freeing oneself" from it. In accordance with other approaches, we talk about a sustainable city when people differentiate their environment from nature but at the same time succeed in peacefully coexisting with it, protecting it from degradation or at least preserving its qualitative features for the coming generations.

6.3 CITIES IN THE DEVELOPING WORLD

In developed countries, the movement of populations from the country-side to the city has mainly been due to industrialization, together with the mechanization of agriculture. More generally, however, the big city with its variety of functions, even in the less industrialized countries, is a rich field for productive and other activities of the urban population. This fact makes the big city attractive; moreover, in many less developed countries with an agriculture that is unable to support the rapidly increasing population, these job opportunities are the cause of a large flow of rural dwellers to the cities. However, employment opportunities for migrants in the urban centres of less developed countries have generally been fewer than anticipated, more so since inadequate education has been an obstacle to their integration and assimilation. Thus the move towards such cities is often a move from the poverty of the village to urban squalor. The vast shantytowns that have mushroomed near big cities in Latin America, Asia, and Africa are examples.

Between 1950 and 1985, the number of people living in cities almost tripled, and the demographic increase continues. These rates have no historical precedent, and the growing urban populations in the developing countries often cannot meet their needs for clean water, hygienic conditions, schools, transport, etc. The phenomenon of shantytowns, with their primitive conditions, overpopulation, diseases and unhealthy environment, is spreading.

The poor in the urban centres of developing countries usually build their dwellings on unsuitable land, such as areas prone to flooding, deserts awash in dust, hills subject to landslides and areas affected by industrial pollution.

They choose these places because their arbitrary dwellings are more likely to be tolerated by the authorities, due to the low price of land in derelict areas. Offsetting this tolerance is a widespread lack of basic infrastructures and great degradation of the urban environment. In some shantytowns of less developed countries, one in four children dies before reaching the age of five and one in two adults suffers from serious intestinal or respiratory ailments. This uncontrolled urban expansion often gobbles up productive agricultural land, with additional negative effects on the ability to provide food for increasing populations.

Any positive prospects for urban development in these big cities are undermined by a lack of financial means as well as the lack of effectiveness of the various measures taken by the authorities. What is more, economic and social conditions are causing cities to expand further and often the policies aimed at discouraging urbanism are cancelled out by other, stronger policies, which make the city more and more attractive to poor people. Thus sustainable development for such cities is a remote objective, as noted by Jorge Hardoy (International Institute for Environment and Development), in 1985:

> Given the distribution of incomes, given the foreseeable availability of resources—national, local, and worldwide—given present technology, and given the present weakness of local government and the lack of interest of national governments in settlement problems, I don't see any solution for the Third World city. Third World cities are and they will increasingly become centres of competition for a plot to be invaded where you can build a shelter, for a room to rent, for a bed in a hospital, for a seat in a school or in a bus, essentially for the fewer stable adequately paid jobs, even for the space in a square or on a sidewalk where you can display and sell your merchandise, on which so many households depend. The people themselves organize and help construct most new housing units in Third World cities and they do so without the assistance from architects, planners, and engineers, nor from local or national governments. Furthermore, in many cases, national and local governments are frequently harassing these groups. The people themselves are becoming increasingly the true builders and designers of Third World cities and quite often the managers of their own districts.

6.4 CITIES IN INDUSTRIAL COUNTRIES

Urban areas in industrial countries are the most important centres for the global consumption of natural resources. They absorb renewable and non-renewable resources and energy from neighbouring or remote areas and have a major impact on the Earth's ecosystems. They are also important loci of industrial production causing a significant part of global pollution. In recent decades such cities have attracted great masses of migrants from poor countries; their economic activity is diverse, with partial de-industrialisation and strengthening of the services sector. This development is creating new problems, but also offers opportunities to reduce pollutant loads and other environmental disturbances in urban areas.

The environment of many industrial cities has improved significantly in recent decades because technologies have improved or polluting activities have been moved elsewhere. Often a reduction in emissions and an improvement in air quality have been noted. In other environmental sectors, however, such as noise, there has rarely been any improvement worth noting. Motor-driven vehicles, especially private cars, are the principal contributors to the environmental burden on urban centres in the industrial world. Nevertheless, the recent slowdown in the upward trend in the number of cars, in conjunction with stricter specifications for exhaust emissions, through cleaner technologies and more effective urban planning, have mitigated the adverse effects of vehicle traffic. Improved public transport, using new, non-polluting, silent and convenient means of transport, plays a decisive part in lifting the burden on the urban environment and improving the mobility of passengers and pedestrians. Pedestrianisation of the historic and commercial centres of many cities has made a decisive contribution to this end.

Public opinion has a vital role to play in promoting changes in the direction of sustainability in cities. In many urban centres, social pressure has caused programmes for mass urban development to be abandoned, imposed a more human scale on residential planning, prevented injudicious demolition of old buildings and historic districts, cancelled programmes to construct urban highways and promoted the conversion of abandoned spaces into recreation areas.

However, many industrial cities suffer from deteriorating infrastructures and social services. Unemployed people, the elderly and minorities are trapped in the vicious cycle of degradation and poverty, as job opportunities and younger, better-educated people abandon the neighbourhoods in decline. The municipal authorities must often manage a heritage that includes poorly designed and poorly maintained dwellings, rising costs and a decreasing tax base. Most industrial countries, however, have the means to address urban and economic deterioration through suitable redistribution policies. In fact, such trends are often reversed through enlightened policies involving cooperation between the public and private sectors, with important investments in staff, institutions and technological innovations. The exodus of the population and of economic activities creates severe economic and social difficulties, but at the same time it makes way for new uses of various abandoned buildings, protects the historic city centres from the threat of demolition and contributes to the renewal of industrial cities.

6.5 NOISE POLLUTION

The problem of urban noise has assumed large dimensions in our times. Sources of noise have increased by leaps and bounds, resulting in more and more areas where noise constitutes a serious disturbance. Characteristic cases are workplaces (factories, building sites, offices), roads, airports, spaces and means of public transport in general, dwellings where there is often little possibility of quiet in the daytime or for much of the night, and entertainment venues, which are often excessively noisy.

Noise is currently one of the most threatening pollutants. It comes mainly from:

- Vehicle traffic, particularly on motorways, as well as traffic congestion
- Aircraft
- Garbage trucks
- Railways
- Motor-driven recreational water craft
- Construction machinery
- Certain industrial activities
- Barking dogs in residential areas
- Entertainment venues
- Some recreational activities inside and outside homes

It is difficult to set limits in this regard, since disturbance from noise is largely subjective. Negative effects on certain animal species have also been noted. Depending on location and activity, people's tolerance varies widely. To a large extent becoming inured to noise is a cultural feature of various societies. There are, however, levels of noise that are unacceptable and harmful for most people. At any rate, the noise-effect relationship is a statistical one. Usually two noise limits are proposed: a level that appears to be endurable to the majority of the population (65–70 dB(A)) and a level that is clearly endurable (55–60 dB(A)). For comparison, we give some typical noise levels in dB(A):

- Auditory threshold (0–10)
- Bedroom (20–30)
- Library (30–40)
- Typical office (50–60)
- Ordinary speech (60–70)
- Motorway (70–80)
- Inside a noisy underground railway (90–100)
- Jet takeoff (110–120)
- Air hammer (120–130)

- Pain threshold (130–140)
- Rocket blastoff (150–160)

The level of 65 dB(A) can be considered an acceptable limit. Above this threshold, when citizens are not suitably protected, noise interferes with the sound of conversations, radio or television to an annoying degree, or disturbs rest and sleep.

Noise has an adverse effect on people's physical and mental health and their quality of life, with a multitude of consequences. Apart from the level, of great importance as far as disturbance and impact on health are concerned is the type of noise, e.g. constant, pulsating, irregular. Auditory acuity may be damaged, leading to temporary and/or permanent loss of hearing. Noise increases stress and arterial pressure. There are indications that cardiovascular damage may be caused by extended exposure to certain types of noise. Even low levels may disrupt sleep and cause insomnia, and also reduce people's concentration, communication and composure. Various nervous or mental disorders, along with lower productivity, may be attributed to exposure to noise.

The problem of noise pollution is extremely important in many countries. Noise is a significant source of disturbance in most large cities. In developed countries, citizens list noise among the main causes of deterioration of the quality of life, and in some cases it is even considered to be the basic cause. All sources of noise are blamed, although means of transport are the most important polluters. Fifty percent of the inhabitants of OECD countries (developed countries) live in areas where the noise level exceeds the comfort threshold, whereas for 15% (1980 and 1999 figures) it also exceeds the maximum tolerance level. More recent data show that the situation has not improved. In some countries it has become much worse in certain "black spots," where dwellings are exposed to noise levels of over 65 dB(A), especially along roads with heavy traffic. In other cases, the "black spots" have recently stopped growing or shrunk slightly, thanks to the use of noise reduction screens or soundproofing in homes. Another trend is the expansion of "grey zones," where the acoustic environment is poor, i.e. in the range of 55–65 dB(A) but has not reached unacceptable levels. Forecasts for the future are pessimistic. A slight improvement in the "black spots" is anticipated, especially around airports, but also a significant enlargement of the "grey spots." The presence of noise is constantly becoming more widespread as periurban and tourist areas are developed, and with increased nighttime activity in cities it is also expanding to fill all 24 hours of each day. In addition, the numbers of vehicles of all types emitting loud noise are increasing. With the emerging prospects for the development and functioning of many cities and the inadequacy of the measures being implemented, the deterioration is expected to continue. Combating noise is generally of low priority in environmental policies, and the outcomes of relevant programmes are often inadequate.

6.6 POLLUTION FROM URBAN SOLID WASTE

Urban solid waste (USW) or refuse (commonly known as rubbish) is produced by dwellings and various services such as offices, shops and the other non-industrial activities of cities. Unlike natural ecosystems, where there are no unusable materials because everything is recycled by decomposers, humans produce large quantities of waste. Prior to the dawn of the industrial age, human societies used most waste products as fertilizers. Waste pollution is intrinsically connected to industrial production, and USW in particular is an important by-product of the activities of a city and often creates serious environmental degradation. Lack of appropriate USW management leads to significant soil, water, air and aesthetic pollution, associated human health problems, as well as an increase in greenhouse gas emissions. Its sustainable management, a challenge that present-day urban societies are called on to meet, should:

• Create no hazards to human health
• Conserve natural resources
• Keep environmental quality high
• Be economically acceptable

Changes in consuming and eating habits, the rapid increase in the quantity of packaging materials that are discarded, and the increase in the population living in urban centres are the main factors contributing to the huge increase in refuse.

In the case of urban wastewater or liquid wastes, mature processing technologies now exist (Chapter 10). The technological solutions available require no social participation apart from citizens' financial contributions towards meeting the operating costs of the installations. Urban solid waste management technologies have not yet reached maturity. The lack of adequate mature technologies has created a need not only for financial input but also for the active participation of generators of USW, i.e. all citizens, in the process of managing it. This necessary social participation consists of an attempt by USW generators to separate waste into different types. This makes their subsequent effective management possible, using technologies capable of valorizing the various products of the sorting at source (Chapter 10) without creating serious environmental problems.

Management of USW includes actions for waste collection, transport, treatment and ultimate disposal. The widely varying chemical composition of USW, the many stages of management, and the urgent need for social participation make integrated USW management a complex environmental and social issue.

Management of USW should be technologically feasible, legally permissible, socially acceptable, and environmentally and economically sustainable. Failure to resolve USW management problems in many developing countries as well as in some developed ones allows the continued existence of unrestricted garbage dumps, piles of rubbish in the streets and the increasing insecurity that rubbish creates for city dwellers. The environmental impacts of lack of management are severe and involve:

- Aesthetic deterioration of the city
- Odour-related nuisances
- Hazards to public health
- Ground pollution
- Pollution of underground and surface water
- Local air pollution
- Greenhouse gas emissions
- Burden on natural ecosystems
- Risk of forest fires
- Economic impacts due to less tourism and degradation of various natural resources, such as coasts

The threats to public health resulting from lack of USW management involve the risk of the spread of pathogens present in contaminated waste, potential nuisances in the form of insects or rodents, the channelling of toxic, carcinogenic, or mutagenic substances into the environment, and the negative psychological effects that the sight of rubbish has on the population. Water pollution is due to toxic inorganic or organic substances, organic material and the products of its decomposition, nitrogen or phosphorus salts etc. Local air pollution is primarily due to emissions of volatile toxic organic substances, while air pollution on the global level is chiefly due to emissions of methane and carbon dioxide (greenhouse gases) from decomposition of USW. The risk of forest fires is mainly due to the possibility of spontaneous combustion of USW that has accumulated near flammable forest vegetation.

The quantity of USW in developed countries generally exceeds 1 and may be as high as 2 kilograms per person per day. Its composition depends to a great extent on the level of economic development and lifestyle. In less developed countries, for example, USW contains less paper, plastic and glass than the USW of cities in the United States, and more organic matter, mainly in the form of food scraps. These differences are decreasing, however, as the gaps between lifestyles are narrowing. The composition of USW generated (before recycling) in the United States in 2006 was: paper and cardboard (34%), organic fermentable, e.g. food waste and garden trimmings (25%), plastics (12%), metals (8%), glass (5%), and other (16%). In South and East Mediterranean countries, the USW contains a higher percentage of organic fermentable (43–70%); a lower amount of paper and cardboard (11–23%) and metals (4–6%); and more or less similar amounts of plastics (11–16%) and glass (2–9%). In Mediterranean countries, USW contains a larger proportion of water (around 37% by weight) as well as a significant proportion of organic carbon (31%) due to the high proportion of compostable substances (biowaste from vegetables, fruits and other foods). USW may contain small concentrations of iron, sulphur, chlorine, lead, zinc, fluoride, chromium, nickel, copper etc. The various metals present in USW, along with the toxic organic substances produced during the decomposition, can seriously pollute the environment, particularly underground water, and also damage health.

Integrated management of USW must include provision for satisfactory solutions at all stages, through optimisation of overall organisation. The stage of final disposal should not be the sole focus of management efforts. Integrated management of USW must aim mainly at minimising the quantity of unusable waste ("rubbish") and valorizing them to the maximum as raw materials or sources of energy. In particular, metals, glass, paper, plastic and construction waste may be recycled and serve as raw materials. Fermentable waste can also become useful when not admixed with other substances. Exploitation requires the separation of USW at the time it is generated (sorting at source); in effect partially eliminating the concept of waste and considering USW as an exploitable raw material from the outset.

Along with sorting at source, integrated management includes an attempt to reduce the quantity and potential hazardousness of the USW generated, in accordance with the principle of environmental impact prevention. The quantity is reduced through efforts to avoid superfluous packaging, make better use of foods, re-use used objects, and in general through reducing unnecessary consumer wastefulness. Potential hazardousness is reduced by avoiding the use of products that contain toxic substances, timely separation of toxic or harmful substances from other USW, etc.

6.7 INDOOR POLLUTION

Particular attention has been paid in the last 25 years to the quality of the air in all types of indoor spaces, such as houses, offices, schools, hospitals, waiting rooms and vehicles. Serious threats to human health have existed since the 19th century in the industrial work environment. There were manifestations of the problem of pollution as far back as prehistoric times, as attested by soot on the walls of caves that served as dwellings. Nowadays indoor air pollution is a serious problem in homes and enclosed spaces in industrial as well as developing countries. Indoor air pollution in the industrial countries is as a rule linked with inadequate ventilation, and has become worse since the 1970s due to improved insulation in buildings. The accumulation of toxic substances in enclosed spaces is often so substantial that the quality of the air is much worse than it is in the external urban environment. The principal indoor pollutants are the following:

- Cigarette smoke
- Radon
- By-products of combustion (carbon monoxide, aromatic hydrocarbons, nitrogen oxides, sulphur dioxide)
- Asbestos fibres
- Formaldehyde and other volatile organic compounds (VOCs)
- Pathogenic microorganisms

Passive smoking is a very important problem, because the smoke from a burning cigarette is just as harmful as or more harmful than what smokers themselves inhale. It contains 3 times as much benzopyrene and 6 times as much toluene (carcinogens), 50 times as much dimethylnitrosamine, and increased quantities of carbon monoxide, acrolein and nitrogen oxides. Cigarette smoke has been proven to contain more than 2000 organic compounds, many of which are carcinogens, poisons, irritants, etc. In most developed countries smoking has been banned in enclosed public spaces such as offices, restaurants, bars etc. in order to avoid the risks of passive smoking.

Radon (Rn) is a radioactive noble gas released from igneous rocks, granites in particular. It is therefore a natural pollutant that may be emitted by floors, construction materials in dwellings, water etc. and accumulates in indoor spaces in concentrations much higher than in the outdoor environment. This gas is chemically inert, but the solid radioactive products of its decay adhere to airborne dust and are inhaled, with the result that they irradiate the lungs and create a threat of cancer. In the United States, radon is the second cause of death from lung cancer, after smoking. In countries where the rocks are mainly sedimentary, the threat from radon is much less serious.

Asbestos is a natural material with unique properties including incombustibility, durability, etc., and has been used extensively as insulation material and in other applications. After years of research, however, it was shown to be a serious threat to health. When inhaled in the form of fibres it causes a form of lung cancer. When it enters the digestive system, however, mainly in water (asbestos water pipes in the water supply system) its potential negative effects are unclear, as it has not been documented to be harmful. The greatest risks are run by workers in asbestos mines and industries, as well as people engaged in demolition or repair of buildings containing asbestos-bearing materials. The European Union has prohibited the production and use of asbestos.

Formaldehyde is outgassed by various insulating materials, glues, furniture, wooden panelling, etc. It is an irritant gas capable of causing nausea or headache, and is a suspected carcinogen.

Pathogenic microorganisms abound in indoor spaces and are spread mainly through central air conditioning systems, air cooling and humidifying units, carpets, rugs, household pets etc.

The easiest and most effective means of combating indoor pollution is frequent ventilation. Suitable care and organisation are also needed, because ventilation decreases energy saving when indoor spaces are heated or air conditioned. Where possible, emission sources (smoking, combustion, formaldehyde etc.) should be avoided. Change in individuals' behaviour plays an important part, and in some cases special air cleaning devices are useful.

6.7.1 Sick Building Syndrome

In many contemporary cities there are a vast number of high-rise glass-walled office buildings with central air conditioning and ventilation. Inside, dense soundproof walls divide the space into small offices into which employees and customers are crowded; these rooms usually contain plastic furniture, synthetic carpeting, photocopiers, decorative plants, many light fixtures, paper files, computers etc.

For several decades now, scientific research on "sick building syndrome" (SBS) has shown that employees in multistory buildings with central heating and ventilation and certain features of internal layout display a variable set of symptoms. They include mainly headache, malaise, eye irritation, nasal congestion, skin irritation, respiratory problems, etc. Surveys carried out in Sweden, Denmark and Great Britain showed that apart from significant damage to workers' health, SBS causes lower performance, loss of working days and substantial healthcare costs.

Modern office spaces provide a sense of luxury and comfort, but they have important disadvantages. Central air conditioning and ventilation are usually designed for a limited number of employees, without taking account of complex interior layouts. When certain limits are exceeded, however, the people, along with furniture, carpets, machinery, stacks of paper, etc., become significant sources of pollution. Air that is not renewed becomes loaded with particles, microorganisms and allergens. Nitrogen oxides and carbon monoxide are also emitted from small electrical appliances; ozone from photocopiers, ionised particles from the radiation produced by old type computer monitors, television screens or light fixtures, along with various VOCs from air fresheners, cleaning fluids, glue, ink, soundproofing materials, etc. Crowded enclosed spaces also may contain large concentrations of moulds, bacteria, viruses and mites, which may cause ailments.

Proper specifications for construction and operation, management, maintenance, and careful cleaning of spaces play an important part. A multitude of studies have proposed measures for hygiene, ventilation and changes in ways of managing and cleaning buildings.

6.8 GREEN AND OPEN SPACES

The existence of green spaces in the city has its roots in the utopian dream urban populations have of regaining their lost contact with nature. An extreme consequence of this tendency would be the abolition of cities, the deurbanisation promised by philosophers and revolutionaries. Proper town planning, including abundant green and open spaces, can be considered a functional compromise, although it entails substantial costs for city dwellers, especially when it is carried out a posteriori. One characteristic example is Paris, where Haussmann, prefect during the reign of Napoleon III, mercilessly tore down thousands of houses between 1853 and 1869 and built 24 squares and broad avenues.

Green and open spaces serve mainly as recreation areas for city dwellers. They make living in cities more pleasant for various reasons, including the following:

- Better microclimate
- Aesthetic enjoyment
- Less noise
- Cleaner air

Pollution reduction in such spaces is due not so much to the vegetation, but rather to lower concentrations of polluting activities, resulting in fewer emissions of pollutants. Moreover, production of oxygen by vegetation in the city is of little significance, since on one hand its quantities are negligible, and on the other city air is renewed via air currents. The concentration of oxygen remains stable (21%) in the air of both cities and countryside; it is related to the stability of the oxygen cycle (Chapter 4). Another important role played by green spaces is their contribution to the presence of wild plants and animals in cities. A substantial number of species of the wild flora and/or fauna adapt to and develop in the built environment and are favoured by the existence of green spaces.

6.8.1 The Periurban Natural Environment

The city is strongly influenced by the periurban natural environment, with which it is in constant interaction. Periurban forests in particular play a decisive role in the functioning of the city, because they may:

- Be important recreation areas
- Play a part in regulating the water cycle

- Hold the soil in place
- Create a milder microclimate

In a city surrounded by mountains, periurban forests offer basic protection from floods, soil erosion, landslides and torrential phenomena.

A substantial proportion of cities have developed in coastal areas. Coasts, as a periurban natural environment, offer cities many advantages, including:

- A milder microclimate
- Rapid renewal of the air
- Direct proximity to an extensive natural ecosystem
- Pleasant location, attractive landscape and seascape
- Increased possibilities for recreation

However, coastal ecosystems require careful management (Chapter 5) because they are vulnerable and can degrade quickly if exposed to thoughtless interventions. In particular pollution by urban wastewater, rubbish, industrial waste and waste from marine transport seriously damages the ecological functions of coasts as well as their image. In addition, careless planning of coastal works and indiscriminate extension of embankments and breakwaters may irreparably destroy the periurban coastal environment.

Urban development, when not carefully planned, has extremely adverse impacts on the periurban natural environment. Tendencies towards chaotic urban expansion are particularly destructive. Fires and clearing of periurban forests are among the consequences of residential pressures on the natural environment. Settlements that are created in this way may be at risk for flooding and torrential phenomena, and also have a poorer microclimate. Seasonal torrents in and around the city need careful management, particularly those whose riverbeds have been filled in, along with illegal construction, culverting and inadequate engineering works, and may create serious dangers to people living in the area.

Important adverse impacts on the natural environment and on periurban forests in particular are also caused by other activities associated with urban development, such as excessive road cutting, modification of major parts of the natural environment to accommodate vehicle traffic and parking, excessive or haphazard creation of recreation and sports venues, plantings of unsuitable and non-indigenous plant species, uncontrolled logging, etc. In general, the harmonious coexistence of the city with its periurban natural environment requires that people and nature be regarded as two separate parts of a single system. Green spaces and the overall natural environment of the urban and periurban space must be regarded as the city's broader ecosystem, the necessary complement to the buildings and other structures, within which the city dweller must adapt and develop as a biological and social being.

6.8.2 Streams in the Cities

An element that enriches the natural environment within the city and at the same time links it to the periurban natural space is rivers or streams with intermittent flow. Stream and river beds in the city:

- Channel off flood waters
- Serve as ventilators that contribute to air renewal in the city
- Are conducive to the development of natural vegetation and fauna
- Form important landscapes of aesthetic and ecological value
- Help create a milder microclimate

Urban activities and interventions must not endanger the unity and natural quality of intermittent streams. This is not only an aesthetic but also a practical necessity, since a stream is able to offer protection from floods when it remains in as natural a state as possible. The core of a stream must be protected, along with a sizeable zone on either side of its banks. Even the areas adjoining the stream must be regarded as a transitional zone and must necessarily constitute part of the overall protection plan. Integrated planning must allow for technical improvements and plantings that stretch along the length of the stream course, from its sources in the mountains to the lowland basins. Planting suitable species on banks that have suffered from human interventions, creating natural filters for debris, and maintaining open streamside spaces in the urban fabric to enable them to act as basins to mitigate flooding are elements that help maintain the stream as a physical entity in harmonious coexistence with the city. In selected places along the stream's course, multiuse elements that will act as poles of attraction of a seasonal nature can be incorporated in such a way as to avoid disturbing the stream's distinctive features and hydrological functionality. Such elements include pedestrian paths, observation posts, small amphitheatre-like spaces, lightweight structures for children to play in, open-air exhibitions, etc. Specified points of entry and pedestrian paths are necessary to ensure the protection of the site's natural quality despite the presence of numerous visitors.

6.8.2.1 Streams in a Mediterranean City: The Case of Athens

Historically, water has been a decisive parameter for the movement of people and goods and the location of settlements. Streams in Greece have always been used as thoroughfares. A French traveller commented in 1853 that:

> The local inhabitants avoided building roads, wherever the water was kind enough to assume this task; in this country, where man seldom runs counter to the work done by nature, streams are kings' highways,

rivulets provincial roads and gullies communal roads; storms have taken up the work of the engineers who build bridges, and the rain, unsupervised, maintains the roads with heavy and light traffic.

The morphology of Attica's land and important natural springs helped to create a large number of streams, of which no more than 10% (70 out of 700) can be identified today. The fertile land near streams has attracted human settlements since ancient times. Streams and rivers were believed to be sacred sites and many sanctuaries have been found along their courses. They were areas for peaceful strolls and recreation. Plato refers to the beauty of the landscape around the Ilissus, where Socrates discussed philosophy with his disciples.

Until around the first quarter of the 19th century, the banks of Attica's streams and rivers retained their character as natural landscapes and recreation areas for the inhabitants of the city. The landscape had not been altered to any great extent and one of its important features was a dense network of streams. Today traces of the old streams can be seen under the broadest roads, especially those divided by a median, which is usually a sign of a culverted stream. However, it is very difficult to trace the entire course of a stream, because streams have been built over along much of their length or have been arranged to form part of the network of subterranean sewer pipelines or part of the road network at ground level. Thus their function as natural spaces and their aesthetic value as natural landscapes have been lost.

In the densely built fabric of the city, where no provision has been made for increased vehicle traffic, even today streams that remain open rather than built over are an inducement to construct new roads, since they are rare linear corridors wide enough to offer great promise as thoroughfares. Thus various types of modifications can be seen, from streams that are completely culverted and given over to traffic to those that have been restricted to a narrow streambed and whose banks are used for traffic. Road corridors follow the courses of the larger streams, whereas the smaller ones have been lost under the urbanized area. When flooding occurs, the unavoidable consequences result.

As regards the periurban space, human interventions have helped destroy the geomorphological character of the basins of streams, woodland flora and fauna, and the natural quality of the areas around streams. Over-grazing, clearing of land, fires, haphazard expansion of quarries and mines, road network which lacks engineering works, use of rubble to fill slopes in order to create building plots: all these things have resulted in increased drainage into built-up spaces. Flood protection works are seldom carried out where they are most needed, i.e. in the part of the periurban space that is farthest upland. They focus mainly on flood protection, through various interventions mainly in the lowland parts of stream courses. Even when streams remain open, their natural quality is altered through overall modifications to the stream-bed and banks using concrete, various types and sizes of structures on the banks, dumping of rubble or wastes, inappropriate plantings, etc. In many cases, a stream cannot be discerned any more as a physical entity.

Overall, in the densely built-up space of Attica one may note the following:

- Intensive exploitation of all properties, land profiteering, and illegal extensions of the city (which in the end are legalised) have caused most of the stream network to disappear.
- Destruction of streams as part of nature has been facilitated by the belief that they are "dirty" places that are hazardous to public health, due to the tradition of disposing of liquid waste and refuse in streams.
- For the most part, the state has treated streams as corridors in the road network, especially when they were the only public lands in areas where roadworks were needed. This attitude, together with the practice of culverting streams and integrating them in the sewer system, has led them to be generally regarded as parts of the city's technical infrastructure.

6.8.3 Nature in the City

The urban space is not made up solely of buildings and human beings, nor is it shaped solely by human activities. Wildlife has always had an important place in the environment of the city. To be sure, the process of building, which transforms a natural or agricultural environment into an urban environment, entails the removal of the flora and fauna from their natural habitats and a great reduction in their populations in the specific area. This removal is temporary, however; the urban system has a tendency to return to its natural state. Wildlife bides its time, ready to rush out and reoccupy urban spaces the moment that human activities cease for one reason or another. But wildlife is also omnipresent during the whole time that humans inhabit the urban space. An important part of it is made up of those biological species that are fed by humans or that generally derive some benefit from coexistence with humans. Various bacteria, both pathogenic and non-pathogenic, fungi, protozoans and viruses thrive and form substantial populations in the crowded inhabitants of cities. Along with this, all sorts of anthropophilic higher organisms thrive, from plants and insects to birds and mammals. Ants, cockroaches, flies, spiders, pigeons, sparrows, rats and mice are examples of anthropophilic animals which usually thrive much better in cities than in the wild.

Depending on available food and suitable spaces where the various species may take refuge and survive, wildlife adapts to a greater or lesser degree to conditions in the city. The most adaptable species are those characterised as anthropophilic, which ultimately form the largest populations in the urban ecosystem. In parallel, however, in certain places or seasons a gradual adaptation of many other species of wildlife can be seen, which begin to frequent or establish themselves in cities. In general, this phenomenon is observed much more with animals than with plants, since vegetation in the city is to a larger extent human controlled, whereas certain animals manage to survive even when persecuted, thanks to their ability to exploit the sites and products of human activities.

The phenomenon of the presence of wildlife in cities has been known for a long time, although it manifests itself in different ways in different historical periods and different geographical entities. The close ties between primitive civilisations and the world of nature allowed many wild animals to live in ancient towns, especially the smaller ones. Pictures and other historical data inform us of the extent of the phenomenon in ancient Egypt, India, China, etc. Many species of wild animals, such as monkeys, are even today plentiful in cities of South and East Asia, are often regarded as sacred and are associated with religion. Cobras often abound even in the heart of densely populated urban areas in tropical Asia. Many cities in Africa even today host wild birds and other animal species. In Europe, the phenomenon has been known for a long time. Medieval cities, for example, were home to many species of birds and mammals, and the use of domestic animals such as horses attracted a myriad of insects and parasites. In medieval London, kites (large buzzard-like birds of prey) and crows were common and were in fact protected by law. Many species of wild flora and fauna benefited from the creation of natural parks within cities. Nowadays urban parks are among the richest sites in the world for wildlife. It is even possible that in the future city gardens and parks will play a vital role in saving many plants that are in danger of extinction.

One proof of the great potential for wildlife to develop within cities was the study of urban areas where human activities had been restricted for some reason. Classic examples were the areas of London, Warsaw and other cities that were bombed during World War II and were very quickly taken over by a wide variety of plants and small animals. However, the rapid expansion of a city and the rapid change in the environment that it brings about, e.g. intensive rebuilding, are factors that make it difficult for wildlife to adapt, and drive it away. Thus, at least in the cities of the developed world, the most and the biggest animals become established where the environment is stable and favourable, e.g. in small, old houses with gardens. In recent years an increased tendency for carnivorous animals and raptors to take up residence in many European, North American and Australian cities has been observed. One characteristic case is that of the fox, which had become established in some suburbs of London by 1950 and spread, mainly in the 1970s, to more than 150 cities in Great Britain. Nowadays foxes have been observed in many other cities, such as Paris, Amsterdam and Copenhagen. No mere visitors to the city, these animals are permanent residents who are born, reproduce, and develop social organisations there. Similar is the presence of coyotes (*Canis latrans*) in US cities, and also wild dogs, which are abandoned domestic dogs that live in many cities all over the world. Birds of prey also occur in many developed urban areas, characteristic examples being hawks and owls, which commonly overwinter in cities, where they feed on the abundant sparrows, pigeons, mice and rats. For some endangered species such as Peregrine falcons, this fact may be decisive for their preservation.

6.8.3.1 The Urban Ecosystem

For the formation of an urban ecosystem, there must be the necessary biological species, suitable habitats, adequate sources of energy, and a rudimentarily stable environment. These preconditions are met in any city, and therefore there is always an urban ecosystem that includes at least people, the parasites that live off them and the microorganisms that decompose dead organic matter. But apart from this minimum food chain, cities, as a rule, also contain other organisms such as plants, animals that feed on refuse, as well as predators of them. Thus a more complex urban ecosystem is formed which includes food chains and primary trophic levels. The various species either arrive in the city from the countryside or are transported from far away by humans. Habitats are formed either in the vestiges of the natural landscape that always remain in a city (e.g. vacant lots, streams), in various structures (roofs, cellars, holes, etc.) or in artificially created vegetation (gardens, parks, etc.). The main source of energy in urban ecosystems, unlike natural ecosystems, is not solar radiation but the introduced food and waste of all kinds. The same sources furnish the various nutrients, since in urban ecosystems recycling of chemical substances (Chapter 4) is incomplete and probably negligible.

A range of important parameters affect the creation, complexity and stability of an urban ecosystem:

- A stable environment, where changes (demolitions, reconstruction, excavations, spatial layout, etc.) are small and rare, is a particularly favourable factor.
- The existence of suitable biotopes in a substantial variety and size. The most important habitat is vacant terrain covered with vegetation. The ratio of vegetation to the area of the city is a decisive parameter. Also of importance are the density of the vegetation, the presence of trees and shrubs, the existence of natural corridors that permit communication between green areas and the short distances between them.
- The natural state of the landscape. The existence of suitable plants, as a rule those that are native to the area, helps create robust, resilient and rich ecosystems. The needs for sustaining vegetation are much smaller when it consists of native plants, and the development of a variety of the area's animals that feed on them is favoured.

Furthermore, the existence in the city of natural formations such as steep rocks, lakes, etc. or even the presence of artificial formations that can play an ecological role (e.g. birds nesting in high buildings) contribute to the variety of the urban ecosystem.

- Abundance of food and water. All sorts of waste and by-products of human activities, stored foodstuffs, etc. are the most important sources of food for many animals, which in turn serve as food for carnivores. One particular and rather important source of food for certain animals in the city and the surrounding area is the dead bodies of animals killed on roads. Water is usually available in cities in sufficient quantities for most wild species. The existence of a high water table constitutes an additional favourable factor for many species of plants.

- Less human-induced disturbance. Wildlife thrives better in cities where people are familiar with nature, and when people tolerate and/or assist it. A positive role is also played by the existence of protected spaces such as parks or other (archaeological sites, cemeteries, hills, river beds, etc.), which may serve a similar function. Air pollution or the presence of waste in general in urban areas are not substantial obstacles to the development of wildlife in the city, although they may injure many organisms. However, the structure and functioning of the city create other negative factors, such as motorways, construction works, the use of toxic or poisonous substances, etc. Automobile accidents are the main cause of death for many species of animals. Thus an urban ecosystem's state also relates to the amount of dangers in each city, as well as with the ways the animals have of protecting themselves.

The status of an urban ecosystem, like that of every natural ecosystem, may be assessed on the basis of fundamental ecological criteria, such as:

- The functioning of trophic levels, and the presence of higher carnivores in particular
- The natural regeneration of vegetation, the normal reproduction of animals, the existence of stages of ecological succession in ecosystems
- The variety of biological species and the stability of populations
- The reduced presence of foreign species of the flora and fauna

Urban ecosystems are a reality in all cities and the areas surrounding them. Wildlife establishes itself in public green spaces and private gardens, as well as various other spots in built-up areas. Urban ecosystems develop to some extent spontaneously in the city, but they can be "fabricated" to a much greater extent. The transformation of the natural or rural environment into an urban environment has important impacts on the functioning of ecosystems. Some of these impacts, however, may be restricted up to a point, or even become positive for wildlife. The parameters of urban eco-systems (stable environment, type and density of vegetation, etc.) hinge on how the city is built and functions. Engineers, urban planners and manag-ers of the urban environment may draw useful lessons from the study of an urban ecosystem, so as to create the conditions for the existence of suitable ecological habitats within the city. A city that has been planned correctly for people is usually more favourable for the development of wildlife.

6.8.3.2 Difficulties Involved in Creating
Urban Biotopes in Some Cities

Opportunities for creating urban habitats in cities do exist, although to a limited extent. One important limiting factor for nature in the city is the disturbance caused by people. In general the reaction of the inhabitants is not always positive for the existence of wildlife in the city. A study in urban parks showed that visitors are interested in clean air and greenery, but not so much in the variety of vegetation or the presence of wild animals. Most important is the insecurity that people feel in parks, particularly at night. Many people complain that the density of vegetation, the lack of electric lighting etc. make parks dangerous. There is therefore a tendency to make green spaces more artificial and therefore less suitable as ecological habitats. In addition, some people are prejudiced against wild animals and are concerned with exaggerated or non-existent hazards. To be sure, real dangers have often been observed from certain animals, such as wild dogs, which attack pedestrians or motorbike riders. To deal with this, proper organisation is required, which may include removal of the animals from such areas. Ultimately, only a very few species, such as songbirds, have gained general favour. Nature-loving city dwellers, who would like green-ery more natural and wildlife closer to them, are in the minority. Urban park security must be improved, a basic precondition for broader social support for urban habitats.

At any rate, overall social and cultural development go hand in hand with environmental friendly awareness, and in the future will probably favour the creation of richer urban ecosystems. An urban environment that is closer to a natural state is gradually becoming more acceptable and desirable. Wildlife in the city helps restore people's contact with nature. The creation of appropriate urban habitats can attract acceptable species and discourage undesirable ones. With the help of proper planning, cities can offer better living conditions in the future for people, along with many species of the flora and fauna.

Chapter 7

Toxic Pollution

DOI: 10.1201/b14609-7

7.1 TOXIC POLLUTANTS, BIOACCUMULATION AND BIOMAGNIFICATION

Many toxic pollutants can be characterised as "general" because they act on various parts of the environment (air, water, ground, biomass in food chains). Their potential hazard for organisms depends at each moment in time on their concentration and toxicity. A number of general pollutants that move by means of foods within the trophic chains appear in increasing concentrations from the lowest to the highest trophic levels. This phenomenon is known as biomagnification and can occur when pollutants are persistent, that is, when they do not break down or break down only with difficulty (e.g. metals, radioactive elements, stable organic compounds) and are not quickly eliminated by organisms. Biomagnification is due to a combination of the low rate at which the pollutants are eliminated or broken down and low energy efficiency all along the food chain (Chapter 4). The concentration of a persistent pollutant on one trophic level may appear on the next level increased by a factor even of the order of 10.

Biomagnification lends another, very serious aspect to some problems of pollution. A pollutant's toxic effect depends on its concentration in the bodies of organisms; therefore even a low concentration of a toxic pollutant in the environment can have a serious toxic effect on an organism if great biomagnification occurs. As a result, the persistence (stability) of a pollutant greatly affects its environmental hazard potential (Table 7.1).

Another important phenomenon that increases the toxic impact of pollutants on organisms is bioaccumulation, which does not occur along a food chain but in an organism's tissues; for example the increased concentration of a toxic pollutant in the fat, the liver, the brain or other organs able to store certain substances.

As a rule, toxic pollution is anthropogenic, because it results directly or indirectly from production activities. However, there are cases where toxic effects are due to natural "pollution." Many organisms produce toxic substances for defence, and many other organisms are susceptible to them.

Table 7.1 List of the 18 Most Dangerous Substances in the Year 2011

2011 Rank	Substance Name	Total Points	2003 Rank
1	Arsenic	1665.5	1
2	Lead	1529.1	2
3	Mercury	1460.9	3
4	Vinyl chloride	1361.1	4
5	Polychlorinated biphenyls	1344.1	5
6	Benzene	1332.0	6
7	Cadmium	1318.7	7
8	Benzo(A)pyrene	1305.7	9
9	Polycyclic aromatic hydrocarbons	1282.3	8
10	Benzo(B)fluoranthene	1252.4	10
11	Chloroform	1207.5	11
12	Aroclor 1260	1190.0	14
13	DDT, P,P'-	1183.0	12
14	Aroclor 1254	1171.7	13
15	Dibenzo(A,H) anthracene	1155.2	15
16	Trichloroethylene	1151.1	16
17	Chromium, hexavalent	1146.6	17
18	Dieldrin	1143.1	18

Source: Agency for Toxic Substances and Disease Register (www.atsdr.cdc.gov/).

Many natural organic substances that occur at high concentrations in food and drink have the potential for long-term toxicity similar to that of synthetic toxic substances.

Next we will examine classes of pollutants that occur frequently in the environment and are associated with serious toxicity phenomena that often accompany pollution. For the most part these are general inorganic or organic pollutants. The risks to humans inherent in many toxic compounds are certainly significant and only partially known. For example, there is still considerable lack of knowledge about the response of the human organism to the vast variety of potentially carcinogenic synthetic substances. In 2006 the European Union published the Registration, Evaluation Authorisation and Restriction of Chemicals (REACH) regulation, on the basis of which a database of chemical substances was created to provide consumers and professionals with information about possible risks. Its objective is to afford better protection of health and the environment through better and timelier identification and notification of the properties of chemical substances. Gradual replacement of the most dangerous substances is also sought when suitable alternatives are found. The different chemical substances in use number over 100,000. REACH will register all the substances produced in Europe or imported in quantities of over 1 ton per year, i.e. around 30,000 substances. There is reason for serious concern about the potential hazard of around 900 of them. Until 2012, 44 substances are proposed to be identified as Substances of Very High Concern (SVHCs) on the basis of their classifications as carcinogenic, mutagenic and/or toxic for reproduction. Five further substances are also proposed to be of great concern. Three of those are respiratory sensitisers and are proposed to be of a very high level of concern for human health. Two further proposals comprise substances proposed to be of a very high concern for the environment due to endocrine disrupting properties.

7.2 VOLATILE ORGANIC COMPOUNDS (VOCS)

These include a great variety of chemicals; many are organic solvents and some have short- or long-term harmful effects on health. They are emitted by many different products, solid or liquid, that are used inside and outside the home, such as paints, polishes, candles, cleaning fluids, insecticides, disinfectants, deodorants, aromatics, building materials, wood preservatives, furniture, office equipment, writing materials, glues, marking pens, photographic supplies, etc. They are more dangerous indoors, where their concentrations may be up to 10 times higher than outdoors.

Effects on health range from irritation of the eyes or the respiratory system, fatigue, allergies, headaches, shortness of breath and nausea to damage to the liver, kidneys and central nervous system or cancer. The effects of exposure to a harmful substance depend on the dose taken up, the duration and manner of exposure, the vulnerability, particularities and habits of the person affected, and the possible presence of other chemicals that may act in synergy with it. The ability of organic chemical substances to have harmful health effects ranges through a broad spectrum from those with high toxicity to those whose impact on health is not known.

7.3 TOXIC SYNTHETIC ORGANIC COMPOUNDS

Thousands of different synthetic organic compounds are in use worldwide and their number is increasing as new products enter the market. Many are dangerous and may have adverse effects on the environment and public health. Some of the most dangerous compounds stand out, as they have three common features: they are persistent (stable) in the environment, they are toxic and they are bioaccumulative. These compounds are known as PBT (Persistent, Bioaccumulative, Toxic) or POP (Persistent Organic Pollutants). They include mainly pesticides, but also additives for cosmetics and plastics and undesirable by-products of certain industrial processes (such as dioxins). Humans and other living organisms come into contact with these dangerous compounds either through direct use of various products, ingestion of food or water, or inhalation.

The United Nations Environment Programme (UNEP) has drawn up a list of the 12 most dangerous categories of organic chemical POP compounds, also known as the "Dirty Dozen." Their common feature is the presence of chlorine, which is associated with the characteristics of stability, toxicity and bioaccumulation. As early as 1982 a broad action network of various ecological and scientific associations was formed, which has begun an international campaign to have the Dirty Dozen banned and replaced by other, less harmful substances, and to introduce the use of alternative plant protection methods. Their production and use were banned by an international agreement among 120 countries signed in Stockholm in 2001. In many developing countries some of these substances are still being marketed or are buried or improperly stored in various places. The Dirty Dozen includes the following categories of substances:

- DDT: An insecticide used widely in the past, and still used in some developing countries. It remains in the soil and in preserved foods for a long time without breaking down.
- Dieldrin: Insecticide used widely in the past.
- Aldrin: Insecticide used widely in the past.
- Endrin: Insecticide.
- Heptachlor: Insecticide, a recognised carcinogen.
- Chlordane: Insecticide.
- Mirex: An insecticide and flame retardant used in plastics, rubber and electronic products.
- Hexachlorobenzene (HCB): A fungicide. Also used in military products, in the manufacture of pigments and other organic materials and as a wood preservative.
- Toxaphene: Insecticide, a mixture of approximately 670 chemical substances.
- Polychlorinated biphenyls (PCBs): Industrial products with many uses; a family of 209 compounds which are extremely toxic and hazardous. They have been used as insulating oil in capacitors and transformers, lubricants, hydraulic fluids, plasticisers, etc. It is estimated that hundreds of thousands of tons have been produced in the past.
- Dioxins: Include 75 extremely toxic substances. They are not industrial products, but by-products of industrial processes or waste incineration (when chlorine is involved).
- Furans: Include 135 toxic substances.

7.3.1 Toxic Substances for the Protection of Agricultural Production

To survive and develop on the planet, humans were and still are forced to alter certain ecosystem balances in order to obtain energy and goods.

Agriculture is such an example that upsets the balance between plants or between plants and animals.

There are millions of species of fungi and other microorganisms, but only around 200 species cause serious damage to agriculture. In nature there is a complex web of relationships between plant and animal organisms and coexistence between parasites and hosts. This balance is to a large extent disturbed by monoculture in large areas, where an unnatural, unstable environment is created. The new conditions of life usually favour the excessive proliferation of one or more microorganisms which harm the specific crop. Without predators or competitors, such parasites will easily predominate and cause great damage, and therefore it becomes necessary to combat them. This is theoretically possible using various biological or physical/chemical means, but in modern-day intensive farming the means used are almost exclusively chemical, the so-called pesticides, which contain various biocides that target parasites, bacteria, fungi, weeds, insects, mites, nematodes, rodents, etc. Certain biocides, such as antibiotics, are also used in intensive livestock farming. Other chemical products used in intensive farming and animal production are various food additives such as hormones.

The biocides are a large group of synthetic organic compounds which are toxic to some categories of organisms and are used in intensive agricultural production as well as in homes and gardens. Their use and applications have grown by leaps and bounds in recent decades and have increased the output of cultivated land many times over, by significantly helping to eliminate harmful insects, parasites and weeds. They may be classed in various chemical groups, depending on the active substance they contain, including the following:

- Chlorinated hydrocarbons
- Organophosphate esters
- Carbamic and aliphatic acids and their esters
- Compounds of chlorotriazines and aminotriazines
- Compounds of the urea group
- Pyrethroids and natural pyrethrins
- Pheromones
- Inorganic salts of As, Zn, Cu etc.

Global industry produces thousands of different pesticide products by combining around 1100 active chemical ingredients. In recent years, with better methods of cultivation and automation, the use of pesticides has increased substantially.

After their application to plants or to the soil, pesticides undergo a series of physical, chemical and biological processes (hydrolysis, oxidation, fission, transport, evaporation, uptake by plants, etc.). They gradually pollute the ground, the water of rivers, lakes and seas, and from time to time appear in dangerous concentrations in foods, in milk and in the fatty parts of the human body. The persistent parasiticides along with the products of their breakdown, which in many cases are even more toxic, move along the food chains and biomagnification occurs. There are also other risks, such as the appearance of adverse effects. For example, in the aqueous environment, parasiticides such as parathion hydrolyse into substances that are capable, in the presence of chlorine, of being transformed into other extremely toxic organic compounds. When such water is used for drinking, serious risks for human health are created. In such cases, the chlorine comes from the artificial chlorination of drinking water for the purpose of protecting public health.

Of particular importance among the biocides are the chlorinated hydro-carbons (DDT, Aldrin, Dieldrin, etc.), which began to be used massively after World War II. Knowledge about them is mainly based on the study of the one most widely used, DDT. The chlorinated hydrocarbons are only slightly soluble in water but highly soluble in fats, and therefore they tend to concentrate rapidly in the bodies of organisms. They are highly mobile and adhere to dust particles and travel along with them over great distances in the atmosphere. If DDT is present in a body of water it escapes along with water vapour into the atmosphere. Chlorinated hydrocarbons have been found in polar ice and in the bodies of animals in the arctic zone.

Chlorinated hydrocarbons are highly stable. DDT has a half-life (the time needed for 50% of its mass to disintegrate) in the environment of around 4 years. Due to the phenomenon of biomagnification, the chlorinated hydro-carbons tend to become concentrated in the higher trophic levels. They act on various organisms in many different ways. In insects, they exert their lethal effect on the central nervous system. Fish appear to be especially susceptible, perhaps because they inhibit the uptake of oxygen through the gills, resulting in death by asphyxiation. The chlorinated hydrocarbons also appear to inhibit photosynthesis in some categories of phytoplankton, something which may have more general ecological impacts. Slower rates of reproduction in birds belonging to higher trophic levels have also been noted. DDT, at high concentrations in the bodies of birds due to biomag-nification, reduces the production of estrogen hormones, with adverse impacts on sexual behaviour and also on eggshell strength due to lack of calcium. Research in this regard has shown that the concentration of DDT in a marine food chain is:

- In marine plants: 0.04 ppm
- In herbivorous fish: 0.23 ppm
- In carnivorous fish: 2.07 ppm
- In higher carnivorous birds: 13.8 ppm

The chlorinated hydrocarbons have a very low direct toxicity to humans and other mammals, but they appear to have adverse long-term impacts.

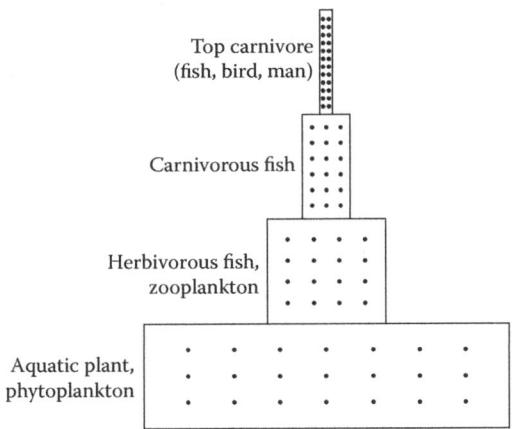

Figure 7.1 Biomagnification of the concentration of persistent pollutants. (From Hadjibiros, K. (2007). *Ecology. Ecosystems and Environmental Protection,* 3rd edition. Symmetria, Athens (in Greek). With permission.)

Large doses in mice have been associated with increased incidence of cancer of the liver. Concentrations of chlorinated hydrocarbons were significantly higher in the bodies of people who died of brain haemorrhage, cirrhosis of the liver and various cancers than in the bodies of people who died of infectious diseases. The patients' histories showed a correlation between concentrations of DDT and the extent to which the patients used it in their homes. In the early 1970s the breast milk of many mothers in the United States contained DDT in concentrations that would have been considered impermissible for commercially sold milk, in accordance with the provisions then in effect.

Bans on the use of chlorinated hydrocarbons in many countries bring about a gradual decrease in concentrations in their human populations, but many years have to go by before there will be enough data on the consequences and before the concentrations become low enough for this pollution to be regarded as harmless (Figure 7.1). Moreover, chlorinated hydrocarbons are still in use in some developing countries, and this is a cause for more general concern due to the high mobility of these compounds.

Another problem caused by the stability of some parasiticides in the soil is that the soil becomes phytotoxic, making crop rotation difficult for years and lowering the quality of agricultural products. This has created serious consequences for exports of agricultural products from countries where excessive use of such substances is common: for instance, countries of the EU have begun to return produce to the supplying farmers due to high levels of parasiticide residues. In general, hazards from pesticides are heightened by irresponsible advertising and lack of information and controls. Even when such substances are banned by some countries, a certain quantity is re-imported in foods from countries where their use is permitted.

Organophosphates, such as parathion and malathion, and carbamates have been used to replace the chlorinated hydrocarbons. Most of them inactivate the enzyme cholinesterase, which is involved in the chemical transmission of nerve impulses. They are characterised by significant to very high toxicity, the organophosphates in particular, but are more or less unstable and thus in principle they do not tend to accumulate, to display biomagnification or to cause long-term damage to organisms. Other serious hazards for people and ecosystems are some insecticides, which poison all living organisms and are capable of penetrating all tissues of plants and animals, killing even the organisms that feed on such plants and animals. Rachel Carson, author of *Silent Spring*, compares them to the magic dress the sorceress Medea, wife of Jason, gave to her rival, who died in excruciating pain.

Failure of workers in agriculture and industry to heed warnings, failure to take protective measures (masks, clothing, provision for wind direction), and high temperatures increase the risks of pesticide use. Poisoning occurs mainly in the following ways:

- Exposed farmers when mixing and applying pesticides, workers on the land who enter fields after pesticides have been applied, and workers in the production and packaging industry are at greatest risk. Even farmers who are well aware of the harmful effects of pesticides are sometimes unable to translate this awareness into their practices. It has been estimated that a high percentage of accidents and poisonings are occupational.
- Accidental ingestion of pesticides by children or other family members and the use of pesticide containers to store or transport food are other means of poisoning.
- Consumption of recently sprayed produce that has been channelled onto the market may lead to poisoning. This phenomenon is common in countries where farmers do not observe the harvest restrictions but pick their produce, mainly fruit and vegetables, immediately after spraying, if better market prices develop.

After the negative aspects of the indiscriminate use of biocides were discovered and following the first ban on chlorinated hydrocarbons (in 1969 in the United States), efforts towards better control of the production and use of pesticides were stepped up internationally, under pressure from groups of scientists and activists.

It can be argued that the organophosphates, carbamates, and even other non-persistent biocides are an indispensable solution to the problem of controlling harmful organisms, provided that they are used with care so as to avoid direct toxic effects on humans and animals. However, as with the chlorinated hydrocarbons, their use has not only direct but also a multitude of indirect consequences. The biocides disturb crucial relationships among species, such as the predator-prey and the parasite-host relationships. They often cause a large decline in the population of predators, followed by a large increase in the prey species that constituted the target population of the biocide. The decline in predators may be caused either directly by the poison or indirectly, due to the decline in the target population and/or other organisms that serve as food for the predators. Furthermore, synthetic biocides activate the mechanism of natural selection, with the result that the hostile organisms develop resistance to their toxic action. With their rapid reproduction, short life cycles and genetic diversification, such organisms have an excellent ability to meet efforts at chemical control head-on and render them ineffective. As the use of pesticides has increased, there has been a rapid increase in the number of species that have developed resistance to one or more pesticides. In 1984 the number of resistant parasites and insects had reached 448, whereas in 1954 it was only 24. The creation of resistant varieties has inevitably led to increased applications and dosage, mainly with more toxic pesticides, resulting in an intensification of the environmental problems.

7.3.2 Non-Toxic Means of Protecting Agricultural Production

It is necessary to control populations of insects and other hostile organisms, particularly in agriculture. But even non-persistent biocides can lead to environmental as well as economic impasses, without taking into account the overall ecological consequences that may be entailed in their widespread, intensive application.

One very interesting alternative is biological control through the introduction of natural enemies (predators or parasites). The implementation of this method requires a careful study of the complex impacts the introduced organisms may have on the environment—the introduction of a predatory animal, the mongoose, to Jamaica to control rats ended in the extinction of most of the small animals in the area. In any event, biological control methods have been developed and improved substantially in recent years. This was aided by the rapid growth of demand for so-called biological products or organic or integrated farming produce. This is a market that has already grown considerably, mainly in developed countries; it promotes the production and consumption of plant and animal products using integrated or benign methods of agricultural production without the use of toxic pesticides or other harmful substances.

There are other methods, more or less well developed, such as artificially increasing plants' resistance to hostile organisms or using biological substances that affect insects' sexual behaviour, reproduction and growth. The use of genetic engineering is another way to achieve protection of agricultural production. Genetically modified plant varieties can resist parasites without the use of pesticides. The opposite is also possible, however, i.e. plants may be genetically modified in order to make them resistant to the use of powerful pesticides that are toxic to ordinary varieties. The prospects for genetically modified organisms are great, and potentially they can reduce environmental degradation from agricultural production; however, they may entail significant environmental hazards and thus they have given rise to strong reservations internationally. They cannot be used safely unless studies capable of providing adequate guarantees are first performed. Their hazards emanate mainly from the following:

- The possibility that various genetically modified organisms with competitive advantages may escape into natural ecosystems
- The possibility that the health of people consuming genetically altered foods may be damaged. To date, however, no such damage has been identified.

The optimal aim of using various methods to protect agricultural production must be to lower to acceptable levels the damage caused by hostile organisms, by minimising ecological disturbances and hazards to human health. In any case, the economic and ecological cost of a drastic treatment that would lead to the complete disappearance of hostile organisms is disproportionately high.

7.4 TOXIC POLLUTION FROM METALS AND INORGANIC SUBSTANCES

7.4.1 Lead

Lead (Pb) enters the human body in a ratio of 2/3 via the peptic system and 1/3 via the respiratory system. Smoking increases its absorption. It accumulates in the organism, mainly in the bones. It impacts the production and functioning of a variety of enzymes, affecting many functions of metabolism. The effects of chronic lead toxicity are far-reaching and are manifested mainly in hypertension, blood composition abnormalities, hyperactivity and brain damage; in children in particular, reduced mental ability has been noted. The effects of lead vary from one person to another. The levels usually measured in human tissues are of the same order of magnitude as the levels that may begin to cause damage, and that is why any further lead burden must be avoided. The US EPA has set a very low limit for drinking water.

It has been proposed that excessive exposure to lead may have been one of the causes of the decline of the Roman Empire. The Romans used lead water pipes and plated their brass cooking and eating utensils with lead in order to avoid copper's unpleasant taste and toxicity. Nowadays lead utensils and lead pipes are not in general use, but industry and cars emit quantities of lead into the environment. The tetraethyl lead added to gasoline to improve its octane rating is significantly more toxic than metallic lead. It is estimated that in the 1970s 400,000 tons of tetraethyl lead were emitted into the air annually, a big part of which wound up in the oceans.

The rate at which lead is mined is around 10 times higher than the rate of natural decay of the rocks containing it, whereas emissions of lead into the air are almost exclusively due to anthropogenic sources. Lead is the only heavy metal whose concentration in the open ocean has increased substantially because of human activities. It has also been found that between the years 1750 and 1960 the deposition of lead on polar glaciers increased 20-fold; however, after 1970 a big decrease was observed. In recent years the more widespread use of unleaded gasoline in cars has significantly reduced lead emissions, resulting in lower concentrations of lead in the air of cities as well as in the atmosphere as a whole.

Plants do not easily take up metallic lead from the soil. However, they take up tetraethyl lead much more easily from the soil and possibly directly from the air as well.

7.4.2 Mercury

Mercury (Hg) is used in the production of chlorine and caustic soda, in the electrical equipment industry, in the manufacture of paints and dental amalgams, in the paper industry, in the manufacture of thermometers, in the production of agricultural fungicides, etc. It is likely that half of the mercury produced annually is released into the biosphere. Another source of mercury pollution is the burning of coal and oil, which contain traces of mercury. However, there are no indications of wide-scale pollution of the biosphere, as has been the case with lead. The presence of mercury in the air is not caused by anthropogenic emissions only but is also due to natural causes, since mercury and its compounds are very volatile. Through the implementation of recycling methods and the restriction of emissions for environmental reasons, the production of mercury fell from 9600 tons in 1971 to 5400 tons in 1989. There is now a trend towards banning the use of mercury thermometers.

Inorganic mercury occurs in the environment either in metallic form, mostly as steam, or in a variety of inorganic compounds. It is toxic but not to a great degree and does not accumulate in tissues. Its principal toxic action is exercised on the kidneys and the liver. The maximum permissible concentration in atmospheric air is 0.05 mg/m^3.

Organic mercury occurs in the form of methyl mercury CH_3Hg^+ (soluble, an important pollutant of water bodies) and dimethyl mercury $(CH_3)_2Hg$ (insoluble). Methyl mercury accumulates in tissues and is particularly toxic. It reaches humans mainly through fish in the diet. It accumulates in the nerve cells, attacks the metabolism of the nervous system and disturbs the action of various enzymes. Its half-life in the human body is around 70 days. The symptoms of poisoning may appear weeks or months after ingestion and include numbness and pain in the lips or hands and feet, ataxia, vision and hearing impairment, and emotional disturbances. This type of illness is known as "Minamata disease" because it appeared for the first time in epidemic form in the small town of Minamata in Japan in the 1950s. There the mercury came from a chemical plant that released its wastes into the sea. It ended up in the bodies of people and animals via the marine food chain. In 1970, 121 serious cases of poisoning including 46 deaths had been officially recognised, but it would appear that the disease affected, to a greater or lesser degree, a much larger number of people. Children born by affected mothers suffered from genetic defects. Other cases of toxic effects have been reported from Iraq, Pakistan and Guatemala, due to consumption of seeds sprayed with a fungicide containing mercury.

In 1969 it was discovered that inorganic mercury can be converted to organic (methyl and dimethyl) mercury by microorganisms that are very common on the beds of seas and other water receptors. In the United States and Canada a problem arose of potential poisoning of humans when unusual concentrations of organic Hg were found in Lake St. Clair's fish. It was believed that this process was responsible for the formation of organic mercury.

The United States and the European Union have set rules for maximum permissible concentrations of mercury in the bodies of fish used for food, and in drinking water, sea water and freshwater.

7.4.3 Cadmium and Zinc

Cadmium (Cd) and zinc (Zn) are metals with similar chemical properties that occur together in nature. Their ecological difference lies in the fact that zinc is a trace element essential for life (the body of an adult human contains 1.4–2.3 grams of zinc), whereas cadmium is not used in biological processes, but is strongly toxic. It has the tendency to displace zinc from many enzymes. Zinc becomes toxic only in very large concentrations, i.e. 10–15 times higher than normal in the organism.

Cadmium is used in metal plating, in the manufacture of plastics, batteries, etc. Substantial amounts are released into the air during the smelting of metals and burning of plastics, as well as during the burning of solid and liquid fuels containing cadmium. The quantities of cadmium and zinc released into the environment from mines are many times greater than those caused by natural erosion. Substantial cadmium pollution is caused by the industrial production of zinc and phosphate fertilisers. The presence of cadmium in the sludge resulting from waste processing reflects industrial and household use of cadmium.

Cadmium is easily taken up by plants and enters their edible parts. For non-smokers, the main agent of entry into the body is food. The rate of entry in relatively unpolluted areas is 30–60 µgr per person per day—the same order as the maximum permissible uptake set by the World Health Organisation. That is why any cadmium burden should be avoided. Additional quantities enter the body through respiration. Smokers run much greater risks, since smoking 20 cigarettes adds 40 µgr of cadmium to the organism.

Cadmium has a long half-life in the human body, and thus it accumulates over time. It mainly affects the nerves and the liver, and some studies have indicated that it may be carcinogenic and teratogenic. The regulations of the European Union have set a maximum permissible concentration in drinking water and bodies of water. The US EPA has set a very low limit for drinking water.

7.4.4 Chromium

Chromium (Cr) can be found in the environment in various forms—as a metal, or as a trivalent or hexavalent ion. In nature it mainly occurs as a trivalent, whereas the metallic and hexavalent forms are produced by industrial processes. Of the chromium existing in the environment, only a small fraction is soluble in water, and fish do not accumulate great quantities of chromium in their bodies. Humans may be exposed to chromium through food, respiration, skin contact in the workplace, ingestion of polluted water or by living near polluted areas. Trivalent chromium is an essential dietary trace element that assists in the metabolism of sugars, proteins and fats. Inhalation or ingestion of high concentrations of hexavalent chromium damages the respiratory or peptic system respectively, as well as the liver and kidneys, and may even cause death; it can also lead to lung cancer. Contact with certain compounds of hexavalent chromium causes skin injuries, and some people are highly allergic to chromium. The US EPA has set a low limit for drinking water.

7.4.5 Nickel

Nickel (Ni) is a metal that is abundant in nature. It forms water-soluble chemical compounds. It is used in alloys for the manufacture of coins, jewelry, batteries and other products. It is released into the air and water from industrial activities. No significant concentrations have been discovered in fish or other animals used as food by humans. Humans are exposed to nickel mainly through food, and also through skin contact with objects such as coins, and to a lesser extent through drinking water, respiration or smoking. The most common adverse effect is an allergic reaction to contact with the metal, since 10–20% of the population are sensitive to nickel. Some sensitised individuals display a reaction when nickel dust is eaten, drunk or breathed in. More substantial exposure is noted in the work environment where more severe symptoms, including cancer, may appear. According to reports from international organisations, some chemical compounds of nickel are carcinogenic to humans, and this may also be true for the metallic form. The US EPA recommends a low limit for drinking water. A limit of 1 mg/m^3 has also been set for the air in the work environment.

7.4.6 Copper

Copper (Cu) is a metal found in nature in various forms, and as a trace element it is essential to maintaining good health. It is released into the environment from mines, industrial and agricultural activity, disposal of liquid waste, and from natural sources such as volcanoes, aeolic erosion, decomposition of organic matter and forest fires. Humans are exposed to copper by inhalation, ingestion in food or drink, skin contact or long-term residence near areas with high concentrations. Drinking water in homes where copper pipes are used may contain important concentrations of copper if the water is acid. High levels of copper that have been absorbed either through respiration or through the peptic system may cause damage. Extremely high doses of copper may injure the liver or kidneys, or even cause death. The US EPA has not listed copper as a carcinogenic substance.

7.4.7 Aluminium

Aluminium (Al) is the most abundant metal in the Earth's crust. In nature it occurs in various chemical compounds; it serves as a raw material for many industrial products. In the environment it occurs in the form of dust in the air, where it may remain suspended for several days; it is also found dissolved in freshwater. It does not occur in significant concentrations in plants or animals. Humans are constantly exposed to low levels of aluminium through food, air, water and the soil. A very small amount enters the body from aluminium household utensils. Higher doses of aluminium are received mainly through inhalation and/or ingestion by people exposed in their work environment or living in areas with high natural or anthropogenic concentrations of aluminium. Low concentrations are not harmful, but high levels, mainly in the work environment, may damage the lungs or the nervous system. Some studies show a link between high concentrations of aluminium and Alzheimer's disease; however, this has not been confirmed conclusively. The limit set for aluminium dust in the air in the work environment is 15 mg/m^3. Household use of aluminium has been deemed safe.

7.4.8 Arsenic

Arsenic (As) is a non-metallic element widely occurring in the Earth's crust. It is dispersed in the air, water and soil, having resulted from erosion of rocks. In the environment it usually occurs in inorganic compounds with oxygen, chlorine and sulphur, but it also forms organic compounds. Inorganic compounds of arsenic are used in various industrial applications, and organic compounds are used as components in pesticides, particularly in the cultivation of cotton. Living organisms such as fish bioaccumulate arsenic, but mostly in an organic form that is not particularly harmful. For humans, its inorganic compounds are the most dangerous. In high concentrations arsenic is lethal. Ingestion of lower concentrations causes nausea and damage to the blood. Inhalation causes irritation of the respiratory system and contact causes irritation of the skin. Long-term exposure to low concentrations of arsenic may cause skin lesions. It has also been listed as a carcinogenic substance. The US EPA has set a very low limit for drinking water, and has banned many of its uses in pesticides. A limit of 10 µg/m^3 has also been set for air in the work environment.

7.4.9 Fluorine Compounds

The toxicity of fluorine compounds has been the subject of public debate due to conflicting opinions on the advisability of artificial fluoridation of drinking water in order to combat dental caries. Fluorine compounds are accumulated in tissues and cause damage to plants, algae and animals. Some plants and microorganisms synthesise toxic organic fluorine compounds. The borderline between dangerous and safe concentrations for humans is not clear-cut, as is also the case with many other toxic materials. Some people are more sensitive than others, and the toxic action of fluorine compounds may be stronger in combination with other pollutants.

A rather wide variety of industrial activities leads to the disposal of fluorine compounds in the air and waters. Thus, during electrolytic production of aluminium, the cryolite (Na_3AlF_6) used as a solvent for Al_2O_3 is in a molten state accompanied by substantial steam pressure. The result is that in the absence of suitable prevention measures, substantial quantities of fluorine compounds escape into the atmosphere, which may fall with rain onto neighbouring land and cause significant damage to livestock farming.

Drinking water is fluoridated at a concentration of around 1 mg/L (as F). There is no convincing evidence that such fluoridation is harmful. The limit in drinking water has been set at 4 mg/L. Scientific documentation of the safety of fluoridation is not complete. Instead of mass fluoridation, there is the possibility of individuals who so desire to apply fluoride simply and cheaply.

7.5 RADIOACTIVITY

7.5.1 Essential Knowledge of the Physics of Radioactivity

Elements whose nuclei undergo automatic transformation, i.e. without the action of an outside force, are known as radioactive. During the transformation, the nucleus emits energy in the form of rapidly moving particles or in the form of electromagnetic radiation. Such emissions cause ionisation of the matter and belong to one of the three types of radioactivity: alpha (α), beta (β) and gamma (γ); even when they consist of particles, they are commonly known as *radioactive radiation*.

Alpha (α) particles are helium nuclei, that is, they are made up of two protons and two neutrons. They are emitted at speeds equal to approximately 10% of the speed of light, and their penetrating power is low. This is because due to their substantial mass, they lose their energy quickly, causing strong ionisation of matter through successive collisions. In the atmosphere they travel for distances of not more than 10 cm or so at ordinary temperatures. As a rule they cannot penetrate human skin (Figure 7.2).

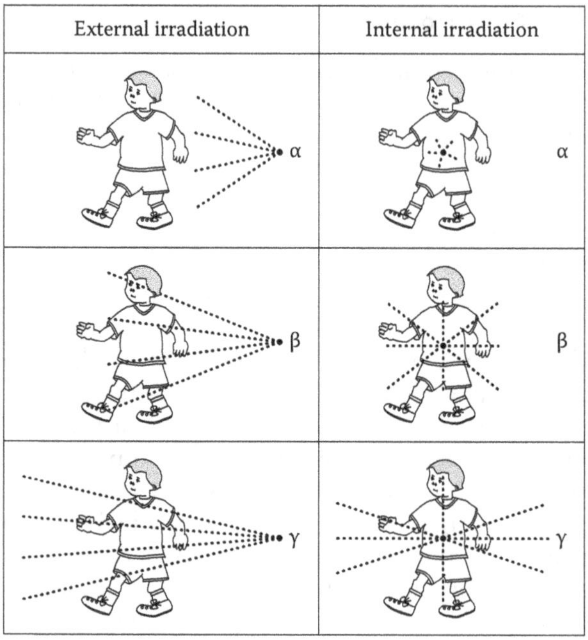

Figure 7.2 Penetrating power of radioactivity α, β, γ.

Beta (β) particles are electrons that result from the fission of a neutron into a proton and an electron, or positrons (positively charged electrons) that result from the fission of a proton into a neutron and a positron. They are emitted by radioactive nuclei at speeds equal to 30–99% of the speed of light. Their penetrating power varies as a function of velocity, but in general it is much greater than that of alpha particles. Beta particles emitted e.g. by the radioactive isotopes Iodine-131, Strontium-90 and Cesium-137 have a kinetic energy of emission much higher than that required for them to penetrate human skin (Figure 7.2), and those emitted by Tritium-3 have a lower one. A sheet of aluminium a few millimetres thick is generally capable of stopping beta radiation.

Gamma (γ) radiation is electromagnetic and is transmitted at the speed of light. It is similar to X (Roentgen) radiation but has shorter wavelengths and therefore greater energy and penetrating power. It travels great distances in the air (Figure 7.2), from tens to hundreds or even more meters. To stop it, a lead wall a few centimetres thick or a concrete wall around one meter thick is needed.

The following units are used to measure emitted or absorbed radioactive radiation:

- One Curie (Ci) is the radioactive radiation emitted by radioactive material that undergoes 37 billion nuclear transformations per second (the transformation rate of 1 gram of Radium-226). The submultiple units millicurie (mCi), microcurie (μCi), nanocurie (nCi) and picocurie (pCi) are used, each of which is one-thousandth of the previous one.
- The Becquerel (Bq) is equal to one nuclear transformation per second, therefore 1 Bq = 27 pCi.
- The rad (Roentgen-absorbtion-dose) is equal to an energy absorption of 100 ergs per gram of irradiated material.
- The rem (rad-equivalent-man) is also a unit of energy absorption, which takes into account its biological effects as a function of the type of radiation and its distribution in the human body. It is the dose that has the biological effects of 1 rad of x-rays and is determined by multiplying rad by a quality factor QF. A unit usually used for smaller doses is the millirem (mrem), i.e. one one-thousandth of a rem.

There is a great variety of natural and artificial isotopes with unstable nuclei that emit radioactive radiation. Isotopes decay due to emission, with a decay rate that decreases exponentially as a function of time. Complete disappearance of the isotope requires a theoretically infinite time. In practice, many isotopes quickly lose practically all their mass and consequently their ability to emit radioactive radiation. It is noted, however, that the radioactivity-induced transformation of the nuclei of an isotope creates in some cases new radioactive nuclei, i.e. a different element, which can emit radioactive radiation with different characteristics.

The decay rate of a radioactive isotope is calculated indirectly using half-life, i.e. the time required for half the original mass of the isotope to be decayed, whereupon the intensity of emission of the radioactive radiation also becomes half of what it was originally. The half-life is of great importance regarding health impacts, because a short half-life means a short lifetime of the isotope but a greater original intensity of radioactivity per unit of mass, whereas a long half-life means the opposite. The intensity of radioactivity corresponds to the number of decays per unit of time and has nothing to do with the kinetic energy of emission of the radioactive radiation, which is given and constant for every radioactive isotope. The half-life is a characteristic measure of each isotope and varies widely. It may be anything from a fraction of a second to billions of years. Frequently encountered isotopes U-238, U-235, Pu-239, Cs-137, Sr-90 and I-131 have half-lives of 4.5 billion years, 710 million years, 24,000 years, 30 years, 28.5 years and 8 days respectively (Table 7.2).

Half-life is different from biological half-life (BHL), i.e. the time a living organism needs to eliminate from its body, through ordinary biological

Table 7.2 Radioisotopes (Radionuclides) That Are Present in Organic Matter or Are Common Products of Fission

Radionuclide		Half-Life	Radioactivity
Argon	Ar^{41}	1827 hours	β^-
Calcium	Ca^{45}	163 days	β^-
Carbon	Ca^{14}	5570 years	β^-
Cesium	CS^{137}	30 years	β^-, γ
Hydrogen (Tritium)	H^3	12 years	β^-
Iodine	I^{131}	8 days	β^-, γ
Iron	Fe^{59}	45 days	β^-, γ
Krypton	Kr^{85}	11 years	β^-
Oxygen	O^{15}	122 seconds	β^+
Phosphorus	P^{32}	14 days	β^-
Plutonium	Pu^{239}	24,000 years	α, γ
Polonium	Po^{210}	138 days	α
Potassium	K^{40}	1.2 billion years	β^-, γ
Potassium	K^{42}	12 hours	β^-, γ
Radium	Ra^{226}	1600 years	α
Radon	Rn^{222}	4 days	α
Sodium	Na^{24}	15 hours	β^-, γ
Strontium	Sr^{90}	29 years	β^-
Uranium	U^{235}	704 million years	α
Uranium	U^{238}	4.5 billion years	α
Xenon	Xe^{133}	5 days	β^-, γ

processes, half the quantity of the absorbed dose of a radioactive substance or other pollutant. Half-life and biological half-life are of great importance when a radioactive substance has entered the organism. The capacity for emitting radioactive radiation is not affected by any chemical reaction undergone by the radioactive substance during the biological processes within the organism or along a food chain. Therefore the only way to get rid of radioactive radiation is a short half-life or rapid elimination of the radioactive substance from the organism. Radioactive substances (but not the radioactive radiation) have the character of persistent pollutants, such as metals and stable toxic organic compounds (Section 7.1), that is, they display bioaccumulation and biomagnification.

The BHL is the same for an element and all its isotopes. A greater BHL means that the radioactive isotope stays longer in the body and therefore more radioactive radiation is absorbed. Uranium (alpha and gamma radiation) has a BHL of 100 days in the human body as a whole, and 300 days in the bones. Plutonium (alpha and gamma radiation) would have a BHL of 200 years, if human life were long enough. Caesium (beta and gamma radiation) has a BHL of 70 days in the body as a whole and 140 days in the muscles. Strontium (beta radiation) has a BHL of 35 years in the body as a whole and 50 years in the bones. Iodine (beta and gamma radiation) has a BHL of 138 days in the body as a whole including the thyroid gland. Tritium (beta radiation) has a BHL of 12 days.

7.5.2 Effects of Radioactive Radiation

A common feature of radioactive radiations is that they cause ionisation of the matter they encounter on their path, hence they are called ionising radiations. The effect on matter caused by alpha, beta or gamma radioactive radiation is due to its energy content. When its kinetic energy is exhausted, the effect ceases. Alpha and beta particles, on their way through matter, collide with the electrons of atoms and knock them out of their orbits, thus creating ionised atoms or radicals as well as free electrons. The free electrons often acquire enough energy from the collision to enable them to then cause ionisation of other atoms. Gamma radiation causes ionisation through various mechanisms, such as excitation of the atoms of the matter through the absorption of the respective energy, resulting in the dislodgement of a peripheral electron.

Ionising radiation causes severe damage to the whole spectrum of organisms. Its effect on living matter and the water it contains results in the formation of a large number of ionised molecules, many of which have high chemical potency. They react with proteins, deactivate enzymes, prevent cell division, destroy cell membranes and in general damage cell function in the short or the long term. Large doses of ionising radiation cause human death shortly after exposure, as was demonstrated in Hiroshima and Nagasaki.

It should be noted that despite their low penetrating power alpha particles are particularly dangerous due to their great ionising action, when the radioactive materials emitting them enter the body with food, water or respiration. In general, the danger of exposure to radioactive radiation depends mainly on:

- The type of radioactive radiation
- The number of fissile nuclei of the radioactive substance
- The half-life
- The distance and the involvement of other materials between the source and the receiver of the radioactive radiation
- The presence of the radioactive substance inside or outside the organism
- The biological half-life (if the radioactive substance has entered the organism)

The effects of ionising radiation on human health are divided into short-term effects caused by large doses received in a short time and long-term effects caused by small doses. The short-term effects of a single dose, like the long-term effects of small doses, vary from one person to another. Typical short-term effects of large doses (average value for the whole human body) are the following: 25 rem is the lowest dose where abnormalities, such as a low white blood cell count, can be discovered through an ordinary blood test. A normal count is restored after a few days. Doses of 100–200 rem have the noticeable effects of vomiting, fatigue, temporary sterility in men and miscarriage in pregnant women. Some deaths are observed at doses of around 200 rem, within about 2 months of exposure. Susceptibility to infections increases rapidly at doses higher than 200 rem. Doses of around 400 rem are fatal for 50% of the population, within about a month. Doses higher than 1000 rem generally cause death within a week.

Humans, like all other organisms, have always received ionising radiation from various natural sources. One category is cosmic radiation, which is mainly galactic or solar in origin. Cosmic radiation occurs on the Earth's surface mainly as secondary radiation caused by the action of cosmic radiation on the nuclei of atmospheric hydrogen and oxygen. The load on the human body is on average 45 mrem/year at sea level and around double that at an elevation of 3000 meters. A second category of natural radiation is that emitted by the various radioactive isotopes in the Earth's crust, which varies widely from one place to another, with a probable average value of 40 mrem/year. In some places of the planet the load is many times higher. A third natural source is the radioactive isotope Potassium-40 in the living tissues of the human body; it entails a load of around 20 mrem/year. With the addition of the other natural radiations, it has been estimated that total natural load in the United States ranges between around 100 and 250 mrem/year, with an average value of around

150 mrem/year. Radiographic and radioscopic testing also cause significant loads, equal on average to 70 mrem/year in the United States, bringing total basic load up to approximately 220 mrem/year.

Humans' basic loads of ionising radiations from natural and artificial sources are certainly the cause of a percentage of cancers and genetic abnormalities; of particular interest, however, is a quantitative estimation of the effects caused by doses of radiation exceeding the basic load. Such estimates have resulted from many experimental studies on animals as well as studies of survivors of Hiroshima and Nagasaki, or of non-cancer patients treated with radiation, or special occupational groups such as workers in uranium mines or children whose mothers had x-rays during pregnancy. The conclusion of the studies is that even small doses of radiation above the basic load cause additional cases of cancer and genetic abnormalities. The so-called linear hypothesis is generally accepted, according to which most of the above-described cases are dose-dependent: that is, if a dose of 0.10 rem causes n cases in a population, then a dose of 1 rem will cause $10n$ cases in the same population.

Regarding the long-term effects of small doses, scientific opinions are divided as concerns the existence of a threshold dose, that is, a maximum dose under which radioactive radiation has no harmful effects. However, the prevailing view is that there is no such maximum value, and therefore there is probably no safe threshold. This means that a small increase in absorbed radiation will bring about a corresponding increase in the cases observed: that is, even very small doses may in the long term cause additional cancers and genetic abnormalities. It has been estimated that in a population of 1,000,000 people a dose of 100 mrem causes 7–30 deaths from cancer, and that the same dose administered to successive generations will result in 6–100 cases of severe genetic abnormalities per million children. These figures are very low in comparison with the "natural" cases of cancer and genetic abnormalities, but they refer to additional cases, i.e. additional lives lost due to radioactive radiation.

Chapter 8

Water Pollution

DOI: 10.1201/b14609-8

8.1 SOURCES AND IMPACTS OF WATER POLLUTION

Urban agglomerations, transport, and industrial and agricultural activities pollute sea, underground and freshwater with their air and solid waste, but primarily their liquid waste. Urban wastewater is carried by sewers and constitutes the bulk of a city's liquid waste. It contains an abundance of microorganisms, organic compounds, certain inorganic salts, mainly salts of ammonium, nitrogen and phosphorus as well as small quantities of other compounds. Marine transport may pollute ecosystems with petroleum products. As a rule, liquid waste from large or small industrial units contain pollutants similar to those in urban wastewater, but sometimes it also contains toxic organic compounds, various metals such as lead and mercury, or radioactive elements. Some industrial wastes damage the aquatic environment because of their high temperature. Agricultural runoff pollutes the aquatic environment mainly with inorganic salts from fertilizers, as well as toxic substances from pesticides. Urban activities also pollute rainwater, which when it runs off city streets is loaded with a wide variety of pollutants such as lead and cadmium, generally in small concentrations. All types of liquid wastes end up in receiving water, which may be on the surface or underground.

The adverse effects of water pollution may be divided into the following categories and involve a wide variety of pollutants. The action of each pollutant may refer to one and/or all the categories of impact:

- Ecological impacts, when pollutants cause disruptions to aquatic ecosystems
- Aesthetic impacts, when pollution causes a disagreeable appearance, smell, colour or murkiness in a body of water (sea, lake, river), degrading the landscape and preventing its use for recreational purposes
- Hygienic impacts, when water (surface or underground) becomes a carrier of pathogenesis or toxicity for the people and animals who either use the water for drinking or swimming or use aquatic organisms for food

Surface waters, i.e. rivers, lakes, springs, lagoons, enclosed marine bays and open seas, are susceptible to pollution to a greater or lesser degree, depending on their capacity for renewal. Here a decisive part is played by the amount of oxygen dissolved in the water. As distinct from the atmosphere, where the concentration of oxygen is high, constant and independent of pollution, water has low oxygen saturation concentrations. Under specific conditions, water is in danger of lacking the oxygen necessary for the respiration of organisms; consequently, the phenomenon of oxygen depletion may occur. The oxygen saturation concentration in water becomes lower as the temperature rises. When there is organic matter in the water, the dissolved oxygen is consumed by aerobic respiration of the microorganisms that decompose it. That is why organic compounds constitute a serious pollutant for water. The consequences may be fatal for a number of aquatic organisms, which are in danger of suffocating. Thus pollution with urban wastewater or other waste containing an organic load may seriously degrade an aquatic ecosystem.

Pollution with inorganic salts containing either nitrogen in the form of nitrates or ammonium salts or phosphorus in the form of phosphates may also have adverse effects on surface water. Such pollutant loads are usually contained in urban wastewater (excrement, detergents, etc.), in certain industrial waste and in agricultural runoff (excess fertilizers). The most important problem created by nitrogen and phosphorus is eutrophication, which consists of an excessive growth of algae (phytoplankton) in surface waters due to an excessive supply of nutrients. This phenomenon constitutes a serious disturbance for an aquatic ecosystem with various adverse consequences including greater turbidity and oxygen depletion.

Groundwater is very susceptible to pollution because of its limited ability to self-purify. The main effect of agricultural runoff and urban wastewater that end up in the underground water table is an increase in concentrations of nitrates, causing groundwater to become toxic to humans and animal organisms. Pollution of groundwater with industrial waste may bring about increased concentrations of metals or other toxic substances.

Another form of pollution of surface and groundwater is contamination, i.e. the presence of pathogenic microorganisms. This is usually a result of drainage of wastewater from cities and from stock-raising activities. Pathogenic microorganisms are detected in water indirectly by measuring fecal coliform bacteria, which are generally not pathogenic; their increased presence is an indication of possible contamination of water with other pathogens.

We now examine in more detail some of the main pollutants of water, their sources and their effects.

8.2 ORGANIC MATTER—OXYGEN DEPLETION

Dissolved oxygen is generally not abundant in water; therefore, organic matter can cause serious pollution, since it can lead to oxygen depletion of water and thus to a severe degradation of an aquatic ecosystem; oxygen depletion may bring about the death of fish or other aquatic organisms and create disagreeable septic conditions through the release of methane and hydrogen sulphide. The oxygen dissolved in the water is not a limiting factor for primary productivity but it may limit populations of aquatic consumers, e.g. fish.

Organic materials are a common pollutant, since wastewater from cities and stock-raising establishments, as well as liquid waste from agricultural and other industries, contain significant concentrations of organic compounds which usually wind up in receiving water. Organic matter serves as food for heterotrophic/chemosynthetic microorganisms, resulting in consumption of the dissolved oxygen in the receiving water due to aerobic respiration, according to the chemical reaction:

$$\text{Organic matter} + O_2 \rightarrow CO_2 + H_2O + \text{heat}.$$

In the technology of water pollution, organic matter is frequently measured by its deoxidising effect, biochemical oxygen demand (BOD). BOD of a quantity of organic matter is the mass of oxygen that is consumed during its biological oxidation; BOD_5 is the mass of oxygen consumed during the first 5 days. In urban wastewater, BOD_5 is usually 65–75% of BOD. For exclusively domestic wastewater, BOD_5 is 55–80 grams per inhabitant per day.

The receiving water is capable of maintaining a maximum concentration of dissolved oxygen, which is known as the saturation concentration. The saturation value depends for the most part on temperature and falls when the temperature rises. Another factor affecting the saturation concentration is the salinity of the receiving water. Due to its greater salinity, seawater has lower saturation concentrations for the same temperature, around 75–80% of the comparable values of freshwater.

When the concentration C of dissolved oxygen in the receiving water is lower than the saturation value, we observe the phenomenon of oxidation during which oxygen from the atmosphere enters and is dissolved in the water, thus enriching it (Figure 8.1).

According to European Union Directives, an absolute minimum dissolved oxygen concentration of 5 mg/L (100% of the time) must be ensured, whereas the average concentrations required are significantly higher depending on the use of the recipient of the wastewater. In the case of marine recipients, significant ecological disturbance may be noted even for values of dissolved oxygen concentration of around 90–95% of the saturation value.

8.3 PHOSPHORUS AND NITROGEN—EUTROPHICATION

Pollution of water bodies with phosphorus or nitrogen salts is caused by overfertilisation of fields or by the disposal of urban or other wastewater.

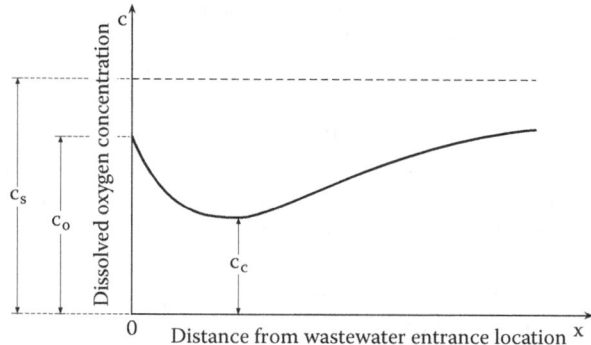

Figure 8.1 Change in the concentration of dissolved O_2. (From Hadjibiros, K. (2007). *Ecology. Ecosystems and Environmental Protection*, 3rd edition. Symmetria, Athens (in Greek). With permission.)

It is important to note that a body of water is not polluted with atoms or molecules of P or N, but with inorganic or organic compounds containing phosphorus or nitrogen. The use of fertilizers in quantities that exceed plant uptake causes excess P and N salts to be carried to surface and underground water bodies via surface runoff and infiltration. Urban and stock-raising wastewater as well as many industrial wastes contain substantial quantities of P and N in inorganic and organic forms. Through the operation of decomposing microorganisms in wastewater treatment plants and in natural ecosystems (aquatic or terrestrial), organic nitrogen undergoes ammonification followed by nitrification, and organic phosphorus is converted into phosphates. Thus the whole P and N content can in principle end up in the ground and in bodies of water, in the form of phosphate anions PO_4^-, ammonium nitrogen NH_4^+ or NH_3 and nitrate anions NO_3^-.

With the usual cations Ca^{++}, Al^{+++}, Fe^{+++}, phosphates form insoluble compounds and this greatly restricts the movement of inorganic phosphorus from the ground towards surface and particularly towards underground bodies of water. This is why pollution of water bodies with phosphorus is usually due not so much to overfertilisation of fields as to discharge of wastewater through outfalls.

For usual pH values, ammonium nitrogen occurs mainly as ammonium NH_4^+, which is held by negative colloids in the soil. In aerobic conditions, however, nitrifying bacteria convert it into NO_3^-. In this form it is mobile and easily pollutes surface and underground bodies of water. Ammonium nitrogen in wastewater can sometimes create a severe problem of oxygen depletion in water bodies during its biological conversion into nitrates, because nitrification is an aerobic process (Table 8.1). For the conversion of 1 gr of ammonium nitrogen into nitrates, 4.57 gr of the oxygen dissolved in the receiving water is consumed. However, nitrifying bacteria

Table 8.1 Factors in Oxygen Depletion of Aquatic Ecosystems

Organic materials	Organic material + O_2 = $\rightarrow$ CO_2 + H_2O: aerobic respiration
Ammonia	NH_3 + O_2 = $\rightarrow$ NO_3 + H_2O: nitrification
Thermal pollution	Decrease in dissolved O_2, increase in metabolic rate of organisms
Eutrophication	Decreased clarity, lack of oxygen, upset of the relationship between photosynthesis and respiration, decomposition of algae

Source: Hadjibiros, K. (2007). *Ecology. Ecosystems and Environmental Protection,* 3rd edition. Symmetria, Athens (in Greek). With permission.

multiply slowly and often the ammonium in the receiving water becomes very diluted before they have multiplied to any great degree.

The most important problem created by nutritive salts, usually salts of phosphorus and nitrogen, is eutrophication, which consists of excessive development of primary productivity, i.e. mainly of algae. Algae are the basic productive organisms in an aquatic ecosystem, and their presence is necessary for it to function. The normal development of algae in an aquatic ecosystem is a natural phenomenon with no adverse impacts. A serious environmental problem is created when there is an excessive proliferation of algae due to the presence of high concentrations of nutritive salts. It should be noted that this phenomenon is a normal environmental state in some shallow lakes or lagoons, where natural eutrophication occurs.

Factors that limit the development of algae may be the intensity of solar radiation or one of the elements necessary for synthesis (C, N, P, etc.); among them, the limiting factor is the factor which is proportionally least available (Chapter 4). In the event that a large quantity of the limiting factor becomes available in the aquatic ecosystem, primary productivity (i.e. essentially the algae population) increases explosively, changing the equilibrium of the ecosystem. There will be a new equilibrium when some other factor becomes a limiting factor.

Sunlight provides the energy needed for photosynthesis (Chapter 2). For most species of algae, solar radiation is not the limiting factor if its intensity is greater than around 2.4×10^4 erg/cm^2/sec. This saturation intensity represents only a low percentage of the intensity of solar radiation, a fact that reveals the low degree of utilization of solar energy during photosynthesis.

The limiting factor in aquatic ecosystems is often one of the elements needed for the synthesis of organic matter, usually nitrogen or phosphorus. Carbon, in the form of dissolved CO_2 and various carbon radicals, is usually present in abundance due to the constant enrichment of the water with atmospheric CO_2.

Phosphorus is usually a limiting factor in freshwater, as is nitrogen in the sea. The biotic community of the algae may achieve adaptations allowing the ratio of N:P that is used to range between 1.35:1 and 13.5:1, in terms of mass. On the other hand, one category of algae, the cyanophyta, are able to fix atmospheric molecular nitrogen; for them, therefore, the limiting factor would be phosphorus or some other element (e.g. iron) or even the intensity of solar radiation.

Unlike what occurs during photosynthesis, during the process of algae respiration O_2 is consumed and CO_2 is liberated into the water. The rate of photosynthesis depends on the intensity of the radiation, which decreases with depth. For a critical value of intensity (about $10 \times 10^2 \, erg/(cm^2sec)$), the rate of photosynthesis of algae is just enough to produce the amount of O_2 consumed in the respiration of both producers and consumers. The depth below the surface at which this phenomenon occurs is called the compensation point. For intensities of radiation lower than the critical intensity, the amount of O_2 dissolved in the water tends to decrease due to predominance of respiration over photosynthesis, and for intensities higher than the critical intensity the water is enriched with O_2. The layer of water above the compensation point is called the euphotic zone. The oxygen produced in this zone exceeds that consumed by the photosynthetic organisms in the aquatic environment.

Eutrophication is natural in some cases. As a rule, however, it is caused by human activities and constitutes a serious disturbance of aquatic ecosystems. Its main adverse consequences are the following:

- Oxygen depletion: Excessive production of organic matter in the form of algae may cause severe oxygen depletion in the receiving water. When there is less sunlight (e.g. at night), photosynthesis decreases, and as a result the demand for O_2 for the respiration of the existing quantity of algae and other organisms is greater than the production of O_2. If the phenomenon lasts for a significant amount of time, it may lead to oxygen depletion. In addition, when the algae die, their protoplasm is used as food by aerobic heterotrophs that consume oxygen. The nutrient elements (mainly N and P) from the dead cells of the algae are liberated into the water, and as a result the phenomenon of eutrophication becomes self-perpetuating. Excessive growth of algae may lead to the creation of a thick layer at the surface. This layer of algae impedes the transfer of atmospheric O_2 to the water (lower oxygenation coefficient) and prevents sunlight from passing through, resulting in a decrease in the photosynthetic activity of the algae underneath. Thus the danger of oxygen depletion increases.
- Less clarity: Excessive concentrations of algae increase water turbidity, and this may have extremely adverse effects on the aesthetic value of recreational waters. This phenomenon is of particular importance for tourist coastal areas that are attractive for their crystal-clear waters.
- Emergence of toxicity: It has been observed that certain (freshwater or seawater) algae may cause illness in humans and animals due to the toxic substances they produce. Cases of poisonings have been

reported, of fish as well as of domestic animals, caused by using water containing high concentrations of algae.

- Accumulation of radioactivity and metals: Algae take up and store toxic metals or radioactive elements that are dissolved in water in very low concentrations. These elements may be relatively harmless for the algae, but through biomagnification (Chapters 4 and 7), high concentrations of radioactive elements or metals may reach humans via fish, agriculture (where water containing algae is used for irrigation), or animals (whose drinking water contains algae).
- Creation of odours, colour and bad taste: Problems of smell and taste in drinking water are generally associated with the presence of algae in sources from which water is taken. The odours and tastes that are created vary according to local conditions and species of algae. Even after raw water undergoes proper treatment, certain odours and tastes remain. High algal growth may also cause colouring of the water.
- Problems during treatment of raw water: When water undergoes filtration in a purification plant, algae (along with other colloidal and suspended materials) gradually clog the filter medium. Chemically assisted sedimentation, using suitable flocculants prior to filtration, may succeed in removing 90–95% of algae. However, the algae that remain may create clogging problems, making it necessary to flush out the filter more often. When algae are present it may be necessary to decrease the amount of time between two successive backflushings (from 30 to 100 hours to less than 10 hours). In extreme cases, the amount of water needed for backflushing may be greater than the amount of water that enters the water supply. Thus the presence of algae slows down the processes of cleaning raw water and increases costs substantially.
- Creation of scum: Slime created by algae is most likely due to decomposition of their cell membranes (a similar phenomenon occurs with bacteria and other organisms). These substances appear in open tanks and open conduits and may create serious problems when the water is used by certain industries (e.g. spotting on paper and food industry products or gelatinous films in liquefaction pipes of industries using cooling systems).

8.4 PATHOGENIC MICROORGANISMS—CONTAMINATION

The most important waterborne pathogenic microorganisms are typhus, paratyphus, dysentery and cholera bacteria, as well as various viruses, primarily infectious hepatitis (with which crustaceans may be contaminated and transmit infection to consumers) and poliomyelitis.

The main source of contamination of water bodies with pathogenic microorganisms is urban and stock-raising wastewater, since it contains human and animal excrement. The pathogens are present in water in very low concentrations and are difficult to detect. For that reason coliform bacteria are used as indicators of contamination. Such bacteria are not generally pathogenic but are excreted by humans in numbers of around 100×10^9 to 400×10^9 per person per day. One group of coliform bacteria, the fecal coliform bacteria, grow only in the digestive system and thus their presence in water indicates pollution with feces and therefore possible pathogenicity. Another feature of coliform bacteria which makes them a suitable measure to indicate pathogenicity is that they are generally harder to eliminate by disinfecting water than are pathogenic bacteria. Thus eliminating coliform bacteria ensures that pathogenic bacteria are also eliminated. The same is not true of viruses, however, because coliform bacteria are not generally a good viral indicator.

Sanitary ordinances set maximum permissible concentrations of coliform bacteria in water according to what it is used for (drinking, swimming, etc.).

8.5 TOXIC WATER POLLUTANTS

From an ecological or hygienic point of view, there are a great number of toxic pollutants. All the important toxic pollutants, such as synthetic organic compounds, lead, mercury, cadmium, nickel, chromium, arsenic, selenium, compounds of fluorine, sulphur, cyanide, the various acids and radioactive materials can be found in water, as well as in the air and the soil. Depending on how toxic they are, they have adverse effects when their concentrations in water exceed certain limits. Pollution may also be transmitted to humans in food through the phenomenon of biomagnification (Chapters 4 and 7), e.g. in fish whose flesh contains high concentrations of mercury that have been increased through the food chain.

Sources of toxic substances are many and varied. Agriculture is the main source of pollution with synthetic organic toxic substances, which are carried to water bodies by surface runoff or by air transfer. Industry is frequently a source of toxic pollutants, which are transported to bodies of water through direct drainage of liquid industrial waste, and gaseous industrial waste, which falls to earth when it rains.

A large proportion of air pollutants originating from means of transport, central heating installations and industry end up in bodies of water. A major part of the lead emitted into the atmosphere ends up in marine ecosystems. Oxides of sulphur and nitrogen are converted in the atmosphere into sulphuric and nitric acid and return to the earth, to terrestrial and aquatic ecosystems, in the form of acid rain (Chapter 9). During the 1980s, pH values lower than 4 were measured in rain over extensive areas of North Europe and North America. Apart from the damage it causes to structures and terrestrial ecosystems, acid rain causes severe damage to aquatic ecosystems. Trout and salmon disappeared from Swedish lakes and watercourses in which the pH fell well below 5.

Chlorine, which is widely used to disinfect water, is a decisive factor in dealing with infectious diseases. In water, however, it reacts with certain organic substances to form some chlorinated organic compounds which may be carcinogenic and teratogenic. Some regulations set the maximum permissible concentration of such compounds in drinking water at 100 μgr/L.

Underground as well as surface water is often loaded with dangerous quantities of nitrates NO_3^-, originating from over-fertilisation of fields or from urban, stock-raising and other wastes that end up in underground and surface bodies of water. In the right conditions, which have been detected mainly in the digestive system of very small children, nitrates are converted biologically into toxic nitrites.

Finally, worth mentioning for its toxic effect on plants, is boron. Concentrations greater than 0.5 mg/L in water used for irrigation cause damage to many plants, and only a few plants are able to withstand concentrations higher than 2 mg/L.

8.6 OIL

In the 1980s, it was estimated that human activities introduced approximately 5 million tonnes of oil per year into the oceans. In a report published in 2002 by the National Research Council (NRC) of the US National Academy of Sciences, the average total worldwide annual release of petroleum (oils) from all known sources to the sea has been estimated at 1.3 million tonnes. However, the range is wide, from a possible 470,000 tonnes to a possible 8.4 million tonnes per year. According to the report, the main categories of sources include natural seeps (46%), discharges (operational discharges from ships and discharges from land-based sources 37%), accidental spills from ships (12%) and oil extraction (3%). Oil may enter the sea through surface runoff, leaks and liquid urban and industrial waste. Another contributor is hydrocarbons that have evaporated on land and fall into the seas along with rain. From the oil pollution due to the marine transport, only a small part is due to oil tanker accidents, with the well-known dramatic local polluting effects. A greater part results from the "normal" operation of many ships and particularly from flushing out ballast consisting of seawater containing a small quantity of oil.

Apart from the well-known, extremely disagreeable effects of oil in the form of tar on swimming beaches, there is a range of other negative impacts. Some of the components of oil are indirectly toxic, whereas others are known carcinogens. Fears have been expressed that the carcinogenic components can affect food chains through aquatic organisms. If hydro-carbons enter the bodies of fish, they can give their flesh an unpleasant taste. Apart from the toxic effect, they damage the feathers of seabirds and may be the main cause of the drastic decrease that has been noted in the populations of certain species of birds. The biological breakdown of oil may also cause oxygen depletion in seawater.

8.7 THERMAL POLLUTION

The warm wastewater from power stations and other industries may cause a rise in the temperature of receiving water which aquatic ecosystems are unable to tolerate. This phenomenon is known as thermal pollution (Chapter 2). In the United States, about 75–82% of thermal pollution is generated by power plants. Some countries require limits on discharges that lead to thermal pollution of receiving waters, although this aspect of water pollution has proven to be more elusive than conventional chemical discharge. Regulation may take different approaches; some laws require a best practice, such as the use of cooling ponds or cooling towers for waste heat discharge. In other cases, a numerical limit on acceptable temperature increase in the receiving waters is applied. For example, the World Bank standard provides a maximum increase of 3°C at the margin of the mixing zone. The US EPA has recommended that temperature increases in receiving water do not exceed 0.8–2.8°C. The lower value refers to the sea in summer and the higher value to watercourses.

Chapter 9

Air Pollution and Global Change

DOI: 10.1201/b14609-9

Air pollution, as well as other environmental disturbances, may be local, regional or global, depending on the area where its consequences are manifested. The air pollution is important for the biosphere because of the vital role of the atmosphere for the environmental equilibrium of the planet and for the organisms. On the other hand, the reversibility of the alterations in the chemical composition of the air is limited. This phenomenon is associated with the partial inability of the atmosphere to assimilate pollution because the self-purifying processes of the atmosphere are only carried out in physicochemical ways. The atmosphere does not contain the functions of an ecosystem, as do the sea or the soil, therefore there are no intricate systemic processes that could increase and accelerate the assimilating capacity. Especially important are the large-scale air pollution phenomena that have a global and non-reversible character, e.g. the increase of greenhouse gases concentrations or the depletion of the stratospheric ozone layer.

9.1 SOURCES AND LOCAL IMPLICATIONS OF AIR POLLUTANTS

Many forms of air pollution are observed in the lower stratums of the troposphere and mainly affect the locations of pollutant emissions, that is, cities and industrial areas.

The main source of air pollution is energy production by burning solid and liquid fuels. The atmosphere of cities and other burdened areas receives pollutants coming mainly from cars, big and small industries, buildings where fuels are used for heating or cooking, construction and other technical works that produce dust and other pollutants, garbage incineration facilities, fluid fuel transport etc. The main air pollutants are:

- Carbon oxides CO and CO_2
- Nitrogen oxides NO_x (N_2O, NO, NO_2 and N_2O_5)

- Sulphur oxides SO_2 and SO_3
- Ozone O_3
- Suspended particulate matter (PM)
- Lead compounds (Chapter 7, Section 7.4)
- Hydrocarbons and other volatile organic compounds (VOCs) (Chapter 7, Section 7.2)

Carbon monoxide (CO) is a colourless, odourless, tasteless and extremely poisonous gas. In the air, it is oxidized to carbon dioxide. It is produced by incomplete burning of carbon at a high temperature with inadequate presence of oxygen. Its appearance in the air of a city is almost exclusively due to exhaust fumes resulting from incomplete burning in the internal combustion engines of cars. It is also a component of cigarette smoke. Its presence in the respirable air in a 0.1% concentration by volume can be fatal. This gas is especially dangerous because it is not easily perceived by the human senses. The first symptoms of poisoning include sleepiness and headache, followed by loss of senses, asphyxia and death. When inhaled, it reacts with haemoglobin, being attracted by it approximately 210 times more strongly than oxygen is, thus substituting oxygen in the blood. Long-term exposure can be a factor for mental retardation or chronic illness. Carbon monoxide contributes to photochemical pollution and it also indirectly amplifies the greenhouse effect because it gets oxidized to carbon dioxide when found in the atmosphere.

Carbon dioxide (CO_2) is produced by burnings and its presence in the air does not harm human health in its usual low concentrations, but in very high concentrations it causes asphyxia. Its accumulation in the atmosphere is the main factor for the amplification of the greenhouse effect.

Nitrogen monoxide (NO) is mainly produced by the oxidation of atmospheric N_2 either in internal combustion engines through the conditions of high temperature and pressure that prevail there or by the burning of natural gas; it is also produced by the burning of nitrogen contained in fossil carbon or petroleum. When in the atmosphere, it is gradually oxidized to nitrogen dioxide (NO_2) which reduces visibility and imparts a characteristic brownish-yellow hue in the air, especially in cities. By further oxidation, nitrogen pentoxide (N_2O_5) may be formed which is an anhydride of nitric acid HNO_3. Generally, these oxides cause several harmful effects in human health. Nitrogen hypoxide (N_2O) occurs in significant concentrations, not specifically in cities but in the whole atmosphere as well, and it is a greenhouse gas. The average global concentration of nitrogen oxides in the atmosphere is only 3×10^{-3} ppm, but in some urban areas it reaches the level of 1–2 ppm.

Gaseous hydrocarbons (HCs) are volatile components of fluid fuels as well as products of incomplete burning. Some are considered to be carcinogenic.

They play an important role in the formation of photochemical smog; CH_4 contributes to the greenhouse effect. They come from the exhaust fumes of cars and from any use or transportation of gasoline, oil and natural gas, due to the escapes that occur.

Sulphur dioxide (SO_2) comes from sulphur contained as an admixture in solid and fluid fuels. It is formed during combustion and is mainly emitted in the atmosphere from industries, diesel-driven cars and heating installations. It has a characteristic odour and contributes to the emergence or aggravation of respiratory system affections. SO_2 can be oxidized in the atmosphere to SO_3 which in turn reacts with the vapours and forms sulphuric acid (H_2SO_4).

Ozone (O_3) is triatomic oxygen that mainly appears in the stratosphere, where it forms a characteristic layer (Section 9.3) and absorbs a part of the incident ultraviolet radiation. It is also found in the troposphere, though, in a very small concentration that increases by pollution and contributes to the amplification of the greenhouse effect (Section 9.4); it also has a toxic action on organisms. It is harmful for plants (80 ppbv reduce the growth rate) and for human beings (120 ppbv cause respiratory problems, such as asthma and bronchitis).

Suspended particulate matter (PM) is solids or liquids found in the air. They are produced by various burnings, by industrial emissions, and by excavations or other activities that create dust. In the atmosphere of cities, suspended particulates are to a great extent produced by the exhaust fumes of cars. Their physical form and chemical composition vary immensely. Some are big enough or dark enough to be visible as smoke or dust. Others are so small that they can only be detected by a microscope. Until 1987, environmental regulations pertained to the total suspended particulate matter, but now attention is focused on the small or very small particulates because they are able to reach deeper into the respiratory system and cause health damage. The EPA limits for PM-10 concentration (particulates with a diameter up to 10 µm) in the air are 50 µg/m^3 (as a yearly average) and 150 µg/m^3 (as a daily average). They can contain a large variety of substances such as various organic materials, lead, cadmium, mercury, and asbestos, many of which are toxic or are carcinogens. The PM-10 consequences on health include damages to the respiratory system and to the lungs, cancer and early death. Elderly people, children and persons suffering from chronic respiratory diseases are more sensitive to suspended particulate matter. Recent studies show that very small particles (with a diameter less than 2.5 µm) are even more dangerous, and there is a trend for the establishment of even lower limits. Suspended particulates are also emitted by natural and various other sources such as volcanic eruptions, agricultural and forest production, aeolian erosion and fires. Suspended particulates can be formed secondarily in the atmosphere through interactions of gaseous pollutants (e.g. SO_2 and NO_x).

9.1.1 Photochemical Smog

High concentrations of nitrogen oxides (especially monoxide) in the atmosphere of some cities, in conjunction with high concentrations of gaseous hydrocarbons and under the influence of solar radiation, lead to a series of photochemical reactions. Solar radiation supplies the energy for the breaking up of chemical bonds of nitrogen oxides and volatile hydrocarbons (or other VOCs). Atomic oxygen and reactive free radicals are produced that subsequently compose a variety of gaseous pollutants. This photochemical smog is a special form of air pollution; its characteristic pollutants are ozone (tropospheric), nitrogen dioxide, and various organic compounds, such as peroxyacyl nitrates (PAN). The presence of photochemical smog is harmful for man—it causes eye and throat irritation and respiratory problems, damage to plants, and reduced visibility.

The presence of temperature inversion that traps and accumulates gaseous pollutants as well as the presence of strong solar radiation are necessary preconditions for the complex reactions that make up photochemical smog. Under normal conditions, temperature decreases higher up from the surface of the earth; therefore the movement of warm air creates upward currents that facilitate the escape of pollutants towards higher stratums. During temperature inversion, a warmer and therefore lighter stratum of air is created at a height of a few hundred meters from the ground, which prevents upward movements and the escape of pollutants; as a result, the trapping of pollutants in the low stratums exposes them to solar radiation for sufficient time for the photochemical reactions to come about.

The basin of Athens (Greece) is a characteristic example of air pollution that started having as its most significant pollutants sulphur dioxide and smoke, and gradually developed into photochemical smog. Non-photochemical pollution, which was high until the 1980s, decreased subsequently thanks to the smoke emissions control, the use of fuels with less sulphur content and the introduction of lead-free gasoline. However, the multitude of circulating cars burdens the basin's atmosphere with great quantities of unburned hydrocarbons and NO_x, and in conjunction with temperature inversion and the strong and long-lasting solar radiation, photochemical smog was often created. During hours of high traffic, high concentrations of smoke and carbon monoxide were observed in some sections of the city.

9.1.2 Effects on Health

Our knowledge in relation to the air pollution phenomena and especially to its effects on human health is insufficient, and even more so in relation to the synergistic action of the pollutants. SO_2, for example, when adsorbed on the surfaces of suspended particulates, can cause serious damage to the respiratory system. The presence of particulates, which offer surfaces that facilitate chemical reactions, accelerates the oxidation of SO_2 to SO_3, which is more toxic. Suspended particulate matter causes health damage in any case, depending on the toxicity of the substances it contains, e.g. lead, mercury, cadmium (Chapter 7), asbestos (Section 6.7) etc.

Research on the effects of air pollution on human health does not easily reach specific conclusions. Conducting realistic experiments on humans is difficult, and inference through observations (e.g. epidemiological studies) is hindered due to the different exposure and sensitivity of different persons and the amount of time needed for the manifestation of illness (e.g. cancer). These difficulties, combined with the complex chemical alterations of atmospheric pollutants and the synergy of their action, have not yet allowed satisfactory knowledge about the relationship between air pollution and human health. Nevertheless, in pollution-burdened cities and industrial areas, air pollution causes—although in a non-dramatic way—a systematic morbidity and mortality increase.

9.2 ACID RAIN

A serious environmental problem, directly connected with air pollution and manifested in a local or regional level, is acid rain. This phenomenon was first reported in industrial areas of Britain in the middle of the 19th century. It has been observed and studied in the past decades mainly in Central and Western Europe as well as in Eastern and Northern America. The great concern that it caused contributed to the establishment of significant measures for the decrease of air pollution (Chapter 10, Section 10.2), resulting in a net amelioration of the situation. However, energy production in many developing industrial countries is based on fossil fuels and especially on coal, which creates preconditions for the emergence of related pollution phenomena in new areas of the planet.

A more correct term for acid rain is *acid deposition*, because acidity can be deposited not only by rain or precipitation but also by gases or dust. The phenomenon is due mainly to anthropogenic emissions of SO_2 and NO_x in the atmosphere that have as a consequence the presence of sulphuric and nitric acid in the raindrops. The pollutants come from industrial areas, urban agglomerations and highways, and they are oxidized in the atmosphere and react with vapours forming sulphuric and nitric acid respectively:

$$SO_2 + O_2 + H_2O \rightarrow H_2SO_4$$

$$NO_x + O_2 + H_2O \rightarrow HNO_3$$

The presence of these acids has as a consequence a smaller or greater reduction of the rain's pH. Acid rain may appear in areas as far as dozens of kilometres away from the site of emission, due to the air transport of pollutants.

The harmful results of acid rain on ecosystems have been documented by a great deal of research. It is known that in many areas of the Earth the soil, due to limestone bedrocks, contains considerable calcium ions that neutralize the increase of acidity through an alkaline reaction. On the rest of the planet the soil, and by extension the freshwater ecosystems, seem to be more vulnerable to the increase of acidity. In lakes and rivers of Northern and Central Europe and Northern America, pH reduction has caused serious damage to many aquatic organisms (Chapter 8). Acid rain has been considered one of the factors that conduce to the degradation of forests of Central Europe and Northern America, mainly because of the lowering of the soil's pH that it causes. The reduction of soil's pH can lead to a downsizing of decomposers' populations or to the release of trivalent aluminium ions that are toxic for many terrestrial and aquatic organisms, although they do not present biological magnification (Chapter 7). Another impact of acid water is the decrease of photosynthetic ability and the capacity for the renewal of foliage of trees. Similar damage may be caused to agricultural cultivations. In coastal marine ecosystems, as for example in the North Sea, the inflow of N through acid rain is a serious eutrophication factor.

In addition to damage to the ecosystems, acid rain can cause significant material damage. It accelerates the erosion of iron, zinc, copper and aluminium resulting in serious damage to many human constructions, especially historical monuments, as well as cars. Especially vulnerable to SO_2 are building materials containing calcium carbonate that turns to calcium sulphate on the surface, according to the reaction:

$$H_2SO_4 + CaCO_3 \rightarrow CaSO_4 + H_2O + CO_2$$

$CaSO_4$ is partially dissolved in water and as a result the stone surfaces of buildings and monuments are destroyed. Affected by the SO_2, the marble monuments of the Parthenon suffered serious damage in the 1960s and 1970s. Therefore they have had to be protected in special closed spaces since the 1980s. The measures for SO_2 pollution reduction implemented since then have prevented aggravation of the problem.

9.3 DEPLETION OF STRATOSPHERIC OZONE

A form of pollution that originates on the surface of the Earth but manifests itself on higher stratums of the atmosphere with consequences that relate to the whole planet is the depletion of the ozone layer (O_3) found in the stratosphere at a height of 20–60 km.

The ozone layer absorbs a great percentage of the sun's ultraviolet radiation and forms a critical shield for the protection of Earth organisms. Absorption at wavelengths 180–240 nm leads to ozone formation from oxygen and at wavelengths 200–320 nm leads to oxygen formation from ozone. Ultraviolet radiation that reaches the atmosphere can be of UV-C, UV-B, or UV-A type. The first contains more energy and is absorbed by ozone to a percentage of more than 99%. Fifty percent of the second is absorbed and only a small percentage of the third, which contains less energy and is relatively harmless, is absorbed.

Strong ultraviolet radiation causes skin cancer, cataracts and genetic alterations to human beings. The DNA absorbs ultraviolet radiation, especially at wavelengths 280–320 nm, and suffers serious damage. The presence of the ozone layer allowed life 400 million years ago to develop out of the oceans and colonize the land. It has been estimated that an increase of incident ultraviolet radiation by 1% would cause an increase of mortality due to melanoma by 0.8–1.5% in the United States. The increase of UV-B in aquatic ecosystems has serious repercussions because of the decrease of phytoplankton. It can also affect terrestrial and aquatic organisms, damage agricultural production and accelerate the decomposition of some plastics. The increase of UV-B that reaches the lower troposphere also causes an increase of the concentrations of tropospheric ozone (Sections 9.1 and 9.4).

Since the late 1970s, scientific observations both from satellites and from the ground have shown a decrease of ozone concentration in the stratosphere. This is especially significant every year in September and October over Antarctica and has become known as the "ozone hole" (Figure 9.1).

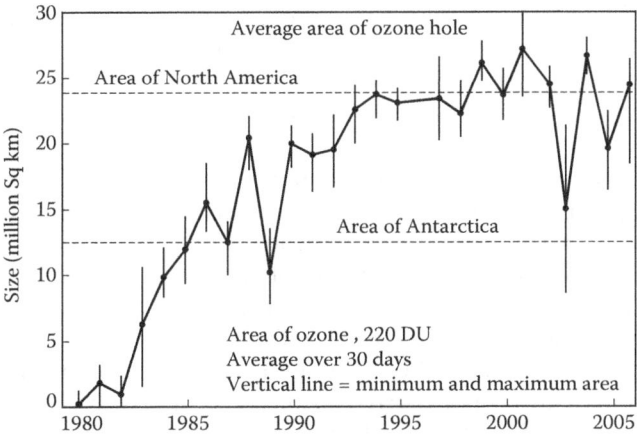

Figure 9.1 The average area of the ozone hole, defined as the region within which ozone is less than 220 DU, is about the size of North America. (From Masters, G.M. and W.P. Ela (2008). *Introduction to Environmental Engineering and Science*, 3rd edition. Pearson, Prentice Hall, Upper Saddle River, NJ.)

Extensive research has shown that there is a significant decrease of the stratospheric ozone layer in many areas of the planet. The quantity of ozone, for example in the middle latitudes of the northern hemisphere, showed a decrease of about 8% between 1980 and 1990.

Since the 1970s, the presence of chlorofluorocarbons (CFCs) in the stratosphere has been considered a principal factor of ozone depletion. Chlorofluorocarbons are not toxic; they are widely used as propellant gases (sprays), as refrigerants (in refrigerators and air conditioners) and as solvents. They are stable compounds in low altitudes but they break down in the stratosphere, releasing Cl atoms that form oxides and other compounds. The chlorine they contain acts as a catalyst and hinders (through Cl oxides) the conversion of oxygen to ozone affecting the reaction towards the left:

$$3\,O_2 \rightleftarrows 2\,O_3$$

It is estimated that a Cl atom that comes from a CFC molecule can break up 100,000 O_3 molecules through successive reactions. Research has shown that chlorine concentration in the atmosphere increased from 2.5 ppbv (parts in a billion by volume) in 1980 to 3.5 ppbv in 1990. Today, specialists suspect that various particulates found in the stratosphere play a non-negligible role in the phenomenon. Such examples are the ice particles of the polar stratospheric clouds and the particles of H_2SO_4 solution observed in the stratosphere. The latter come from industrial emissions, biological processes and volcanic eruptions. These particles do not destroy ozone directly but they reinforce the action of chloride compounds produced by the breakup of CFCs.

Recent scientific papers, without questioning the primary role of chlorine, suggest additional mechanisms of ozone destruction through, for example, bromine compounds or organic compounds. Also, the role of nitrogen oxides in the stratosphere has not been fully clarified. The intricate chemistry of the stratosphere is not yet sufficiently known and there are probably other important mechanisms that have not been identified. The atmosphere exhibits a high vulnerability to changes of its composition, partially because it is not able to assimilate pollution through ecological processes. Nevertheless, the CFCs, in addition to their role in ozone destruction, they are considered significant greenhouse gases (Section 9.4).

The most drastic CFCs are CFC-11, CFC-12 and CFC-113. The damage caused by the emissions of CFCs dictate their substitution by other chemical compounds that do not destroy ozone. The dramatic character of the ozone hole and the fear that the depletion of the stratospheric ozone layer could be faster than what was predicted by the mathematical models drove the international community to respond immediately to the scientists' concerns. The international agreements signed for the protection of stratospheric ozone (Vienna Convention in 1985, Montreal Protocol in 1987 and five revisions of it) prescribed the imposition of limitations and gradual abolition of the production and use of the most dangerous CFCs by 1996, with a provision for a total abolition of the less active ones until 2030. Some developing countries delayed signing the Montreal Protocol and continued using CFCs for the production of refrigerators and air conditioners for some years. Nevertheless, the international cooperation that led to the rapid confrontation of the problem is considered to be a model one. Some scientific research has already shown an improvement of the ozone condition, although a complete recovery should not be expected before the end of the 21st century.

9.4 THE GREENHOUSE EFFECT

9.4.1 Composition and Thermal Equilibrium of the Atmosphere

Although it ranges among the extremely low temperatures of space, Earth's surface maintains a temperature that is mild and favourable to life, with a mean value of about 15°C. Its direct warming by the sun, along with a small contribution of heat from the planet's interior, justify an average temperature not higher than −18°C. The observed positive difference of 33°C is due to the natural process called the *greenhouse effect* which warms the lower atmosphere (troposphere) and conduces to the creation of an environment that is friendlier to life. Contributors to the phenomenon are some trace gases of the Earth's atmosphere that are usually referred to as greenhouse gases. They are relatively transparent for the incident visible and ultraviolet solar radiation but they absorb infrared radiation; as a result, they trap, like a greenhouse, the infrared radiation emitted from the surface of the Earth. The Earth is warmed by the whole spectrum of the incident solar radiation and it emits infrared radiation (according to the Stefan-Boltzman law, Chapter 2, Section 2.7); consequently it achieves a state of thermal equilibrium. The presence of the greenhouse effect results in the increase of the average temperature of the planet's thermal equilibrium by 33°C.

An increase of the concentrations of greenhouse gases in the atmosphere causes a change in the radiation equilibrium and a concomitant disturbance of the existing thermal equilibrium due to an increase of the energy quantity that is trapped in the troposphere. That entails more violent atmospheric phenomena (storms, typhoons etc.), an increase of the average temperature of the atmosphere and the oceans and generally changes to the planet's climate. There are strong indications that air pollution caused by human activities amplifies the concentrations of greenhouse gases and therefore the intensity of the greenhouse effect which, whereas in the past was a

Table 9.1 Greenhouse Gases That Are Affected by Human Activities and Which Contribute to the Greenhouse Effect

	CO_2 (carbon dioxide)	CH_4 (methane)	N_2O (nitrous oxide)	CFC-11 (chlorofluoro-carbon-11)	CF_4 (perfluoro-methane)
Pre-industrial concentration	~280 ppm	~700 ppb	~270 ppb	zero	40 ppt
Concentration in 1998	365 ppm	1745 ppb	314 ppb	268 ppt	80 ppt
Rate of concentration change	1.5 ppm/yr	7.0 ppb/yr	0.8 ppb/yr	−1.4 ppt/yr	1 ppt/yr
Atmospheric lifetime	5 to 200 yr	12 yr	114 yr	45 yr	>50.000 yr

Source: M. Bell and J.C. Walker. 2005. *Late Quaternary Environmental Change* (2nd edition). Pearson Education, London. With permission.

natural phenomenon, now has become the most serious environmental problem (Table 9.1).

The most important greenhouse gas is water vapour (H_2O). However, the role of vapours in the warming of the atmosphere is double and complex because they amplify the greenhouse effect but also form clouds that, by increasing the reflection of solar radiation towards space, cool the atmosphere. Moreover, human activity during the industrial era did not exert great influence on the average humidity of the atmosphere; therefore it did not exert much influence on the quantity of water vapours. Thus, among the greenhouse gases whose increase is anthropogenic, CO_2 comes first (a portion of more than 60%), CH_4 (a portion of more than 25%), CFCs (a portion that is decreasing by virtue of the Montreal Protocol), N_2O and the troposphere's O_3 follow. Energy use in industry (24%), energy use in buildings (17%), land use changes, forestry and agriculture (18%), transport (between 15% and 20%), fugitive emissions and waste (9%) are the main anthropogenic activities that amplify the phenomenon.

The increase of CO_2 concentration (Figure 9.2) is due to the biogeochemical carbon cycle perturbation (Chapter 4, Section 4.3.2) from the use of fossil fuels and deforestation. It is estimated that CO_2 concentration was about 290 ppm in the middle of the 19th century; it reached 415 ppm in 2021. The emitted quantities of anthropogenic CO_2 are in fact about three times higher than the ones estimated on the basis of the observed increase of its concentration in the atmosphere; however, they are absorbed by approximately two-thirds, mainly by the ocean and by land vegetation. All kinds of fuel burning contribute to the warming of the atmosphere only indirectly through CO_2 emission, while direct warming of the atmosphere due to heat emission is negligible. The same is true for heat emissions by the ecosystems.

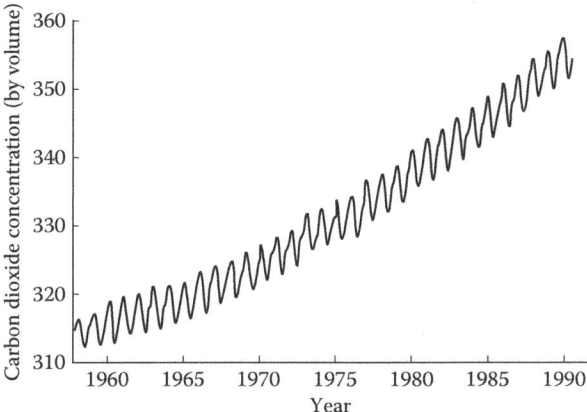

Figure 9.2 Biological rhythm in the atmosphere. Concentration of the comparatively rare gas carbon dioxide fluctuates annually, rising in the northern winter with respiration in leafless landscapes, falling in the northern summer as photosynthesis draws down the atmospheric supply. The measurements are from Hawaii, where the lack of local industry allows measures of unpolluted Pacific Ocean air. The global carbon pulse registers events in the northern hemisphere because most land, and hence most plant production, is in the north. The sinister upwards trend in the data is the best evidence that CO_2 concentration is rising as a result of burning fossil fuels and clearing forests. (Colinvaux, P. (1993). *Ecology* 2. Wiley, New York. With permission.)

CFCs have an exclusively anthropogenic origin; they did not exist in the atmosphere before 1930. They are considered responsible for chemical reactions in the stratosphere that cause depletion of the ozone layer (Section 9.3) but they also act in the troposphere as potent greenhouse gases (Figure 9.3).

The concentration of CH_4 in the atmosphere began to increase at the outset of the industrial era and today it has more than doubled. Its main sources are considered to be every kind of anaerobic process (Chapter 2, Section 2.5), especially from cattle raising, rice cultivation and landfills (Chapters 6 and 10) as well as mining and use of fossil fuels, mainly natural gas.

Nitrogen oxides (especially N_2O) as well as tropospheric O_3 are considered to be greenhouse gases of secondary importance. The increase of N_2O is mainly attributed to the nitrogenous fertilizers and perhaps to the use of fossil fuels. The increase of tropospheric ozone is due to the air pollution of the cities, forest fires, increased UV radiation due to depletion of the ozone layer and possibly other causes.

The relationship between the greenhouse effect and the depletion of the O_3 layer is not limited to the action of CFCs that aggravates both these problems. The large-scale perturbations of the physicochemical conditions of the stratosphere are connected in various ways, and relevant knowledge is still insufficient. For example, the intensity increase of UV radiation near

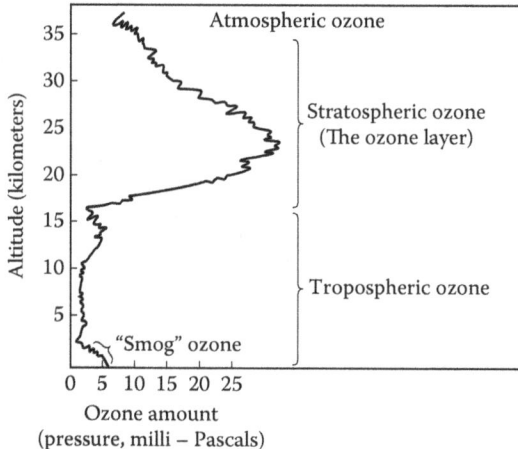

Figure 9.3 Amounts of ozone in the troposphere (10% of atmospheric ozone) and stratosphere (90% of atmospheric ozone). Ozone in the stratosphere plays a beneficial role: it acts as a primary UV radiation shield; current issues: long-term global downward trends, springtime Antarctic ozone hole each year. Ozone in the troposphere plays a harmful role: it has toxic effects on humans and vegetation; current issues: episodes of high surface ozone in urban and rural areas. (Modified from Masters, G.M. and W.P. Ela (2008). *Introduction to Environmental Engineering and Science*, 3rd edition. Pearson, Prentice Hall, Upper Saddle River, NJ.)

the surface of the Earth entails an increase of the tropospheric O_3 as well as a decrease of the sea algae photosynthesis. Both these changes entail an amplification of the greenhouse effect.

In the beginning of the 1990s, a significant deceleration of CO_2, CH_4, and CFC's increase rate was observed. For the first two gases, a probable cause is considered to be the great eruption of the volcano Pinatubo in the Philippines in June 1991; it supplied large quantities of SO_2 to the stratosphere causing significant physicochemical changes. Also, the improvement of the control of the natural gas escapes has probably contributed to the reduction of CH_4 increase.

The significant reduction of CFC's increase rate observed today is undoubtedly due to the international agreements for their reduction, which decelerate amplification of the greenhouse effect. Reduction of CH_4 emissions is also an important issue. However, long-term control of CO_2 increase remains the greatest challenge for dealing with the greenhouse effect. Reducing energy needs, promoting material recycling and CO_2 storage in geological formations (CCS, carbon capture and storage) would give positive results. The issue of energy production is decisive, especially that of electricity production and energy saving. Renewable energy sources are a promising alternative that offers environmental security and can cover a substantial part of the energy deficit that would be caused by the reduction of fossil fuels. In the array of various renewable energy sources, wind energy is a mature technology and is the most promising solution, followed by exploitation of solar energy, utilization of biomass and agricultural waste, and wave and tidal energy. Hydroelectric energy is also a renewable source but it may have potentially high environmental impacts. Solar energy production is growing rapidly in some countries due to the increase of the use of photovoltaic systems and other technologies. Key concerns are how much it will cost to "stabilise" the climate and minimize manmade climate change and to what extent the cost of avoiding dangerous climate change will compete with other priorities such as development.

9.4.2 Earth's Climate

The alterations brought about by human activities on the environment during the industrial era, the principal one being air pollution but also land use changes, are now manifested on such scale that they influence the general state of the planet. They are considered on the whole to be an interrelated system of disturbances usually called *global change*; the intensifying greenhouse effect and the attendant anthropogenic climate change (which, according to many indications, has already begun), as well as desertification and perturbation of the hydrological cycle, are the most important.

The climate is the total of the average values of various elements that define the weather, such as temperature, humidity, rains, winds, snow as well as violent meteorological phenomena like storms, typhoons etc. All of these are manifestations of the movements and changes of the troposphere that interact with the various physicochemical processes, the action of live organisms and extraordinary events like volcanic explosions. The driving force of all this is solar energy that gets trapped in the atmosphere. Consequently, the amplification of the greenhouse effect, which means more energy in the atmosphere, entails not only a rise in temperature but also a wider perturbation of the planet's climate.

The climate of the Earth does not present great changes mainly because of the long-standing stability of the atmosphere. A different composition of the atmosphere would probably mean a very different average climate. Of course, the climate presents great differences according to the geographic latitude and local geomorphological conditions but it also undergoes small changes in relation to time. Thus the climates of various regions have undergone observable changes throughout history. Man has always been adapting to the climate of his region, and sometimes he has succeeded in changing it on a local level.

Various paleontological findings and other scientific data inform us about Earth's climate in the last million years of its geological history. There is a continuous cyclic alternation between a warmer and a colder climate, with a basic period of about 100,000 years, one part of which corresponds to a warm period with climatic conditions similar to the ones we have today, while the rest corresponds to a cold glacial period. As the Earth rotates around its axis and around the sun, various periodic variations occur that bring about changes to the intensity of solar radiation which falls on every spot of the planet. The variations are due to gravitational influences of other planets, tides etc. The change in the eccentricity of the Earth's orbit has taken place in a period between 90,000 and 100,000 years. The change in the obliquity of the ecliptic, or axial, tilt has taken place in a period of about 40,000 years. The Earth's orbit tilt changes with a period of approximately 70,000 years. The direction change of the rotation axis has a period of about 20,000 years. The Milankovitch cycles are the collective outcome of the influence exerted on the climate by the changes of the Earth's movements. However, the correlation between astronomic variations and climatic changes cannot be considered complete for the time being. Today, approximately 10,000 years since the last glacial period, we are going through a warm period, the duration of which there is no satisfactory forecast. It is probable, on a geological scale, that the planet is approaching a new glacial era. It has been calculated that in about 30,000 years, Earth's orbit will have changed enough to reduce sunlight in the Northern Hemisphere to the levels that led to the last ice age. In any case, the study of the climate's course during the last 130,000 years allows the representation of the conditions that prevailed at the end of the previous warm period and the phenomena that followed, as for example the gradual covering of North Europe with ice, the gradual replacement of South Europe's forests by steppes etc.

In addition to the fluctuation that corresponds to the great climatic cycle of the last 100,000 years, it has been ascertained that the climate during the post-glacial period presented a series of smaller fluctuations for which there are no satisfactory explanations. We are talking about an alternation of warm-rainy periods and cold-dry ones, the average temperatures of which did not differ by more than 2°C. Those differences, however, were enough to significantly influence the vegetation, fauna, soil and waters, which had an important impact on the corresponding historical periods. According to certain views, the warm periods were eras of flourish and economic prosperity for humanity, while the cold ones brought years of misery and economic retardation. According to these views, the era of prosperity of primitive peoples that corresponds to the myth of Eden, the decline of great kingdoms in Mesopotamia, Egypt and the Indies, the development and amassment of capital during the Middle Ages in Europe and North Africa (Medieval Warm Period) or certain phenomena of economic deceleration and retardation that were observed during the period called Little Ice Age (about 1450–1850) could be explained.

9.4.3 The Danger of Climatic Change

Monitoring of the temperature since the middle of the 19th century shows that the average temperature of the planet has increased by approximately 0.5°C in the last 100 years. This change is compatible with the warming that would be caused by the observed increase of greenhouse gases during the same period. However, this finding is not enough to prove that global warming due to intensifying greenhouse effect has already been manifested, because for the time being, the alterations are still within the limits of "climatic noise" and could be attributed to natural fluctuation.

The manifestation of an intensifying greenhouse effect during the coming decades is a serious scientific prediction. There are, however, great uncertainties in relation to the changes that it will bring about both to the average climate of the planet and—even more so—to the local climates of different geographic regions. The main factors of uncertainty that prevent a more accurate prediction about the expected climatic change are:

- The future course of gases' concentrations and the ensuing strengthening of the greenhouse effect
- The climatic system's sensitivity, i.e. the degree of its response to the increase of the quantity of trapped radiation in the atmosphere
- The "thermal inertia" of the sea, which decelerates the rate of atmosphere warming

Most of the greenhouse gases will probably have increased concentrations in the atmosphere of the coming decades. However, forecasts are uncertain due to insufficient knowledge about biogeochemical cycles, but mainly about the future policies for energy, agriculture and the forests. Increasing CO_2 emissions are probable, especially in developing and new industrial countries, because of increased coal and natural gas usage, although decreased petroleum usage is expected due to increased cost and decreasing supply. It is expected that there will be a medium-term reduction of CFC concentrations through the application of international agreements for the protection of the ozone layer. The predictions usually refer to the "equivalent" CO_2 concentration in which the contributions of the rest of greenhouse gases are included. A symbolic date is that at which "equivalent" CO_2 will be doubled in relation to its concentration before the industrial era. Estimations place this symbolic event at around 2050.

The estimation of the climate's sensitivity is carried out through the general circulation models (GCMs) of the atmosphere coupled with the ocean and the vegetation, which are simplified representations of the climatic system's operation and "run" on the biggest modern computers. Based on these techniques, it is predicted that the doubling of equivalent CO_2 will cause an increase of Earth's average temperature by 1.5°C to 4.5°C. The limited accuracy of this conclusion reflects the existing uncertainty in the parameters and calculations. In relation to predictions about the rest of the climatic changes (rainfall, storms etc.) the uncertainty is even greater and the same is true for predictions about the distribution of changes in various regions of the planet. It is considered, though, that atmospheric precipitations (rain, snow, dew etc.) will increase as a whole but their distribution will become more uneven and that the temperature will rise more in places and seasons (high latitudes, winters) where it is lower, than where (equator, summers) it is higher.

The climatic parameters' changes occur gradually and with a delay in relation to the greenhouse gases' increase. Positive and negative feedback will play an important role; feedback includes future atmospheric concentration of water vapour which is a powerful greenhouse gas, the complex impact of cloud radiation, the ocean circulation that plays a large part in determining climate because of large heat capacity and heat circulation as well as ice-albedo because ice and snow are powerful reflectors of solar radiation.

The completion of the parameters' response to a given increase probably requires some decades, because a part of the additional heat is absorbed by the ocean, and the slow process of its warming delays the final warming of the atmosphere. Thus the climatic system takes a long time to reach an equilibrium and there is always a significant difference between the observed climatic alteration (temporary) and the one to which the system is subjected (permanent, since it corresponds to equilibrium) at the same moment in time. For example, the planet is subjected to a temperature increase of 0.3–1.9°C if the greenhouse gases remain stable at 1990 values. Respectively, the temperature change that will be observed at the date of the equivalent CO_2 doubling (a total rise of 1.5–4.5°C for the state of equilibrium) will range between 0.5 and 2.5°C. Of course, this equilibrium is theoretical because each new increase of the gases entails alterations. For the climatic system to achieve equilibrium, the greenhouse gases' concentrations should be stabilized and sufficient time should pass so that the warming procedure of the pair sea-atmosphere may be completed.

9.4.4 Possible Impacts of Climatic Changes

Temperature increase is only one of the impacts that will follow a potential warming of the planet. It is true that the alteration will be gradual, but in a pace unprecedented for geological givens. The temperature increase, but even more so the alterations of the rest of climatic parameters, especially that of the hydrological cycle, will have a dissimilar distribution on a global level.

One of the most direct and large impacts is the increase of violent meteorological phenomena both in intensity and frequency. The accumulation of great quantities of energy in the troposphere, due to trapping of radiation, will be defused by typhoons, gales, storms etc. The economic side of these threats has already caused major worries to insurance companies, inter alia, that will be called to cover the damages.

One consequence that is expected with certainty is the rise of the sea level, which will be due first to the thermal expansion of water and second to the melting of glaciers. Many mountain and arctic glaciers have already undergone significant reduction. The ice cap of Antarctica may not be reduced to a great extent during the first decades of change because of the increased precipitation (snow) that it will receive in a context of global warming. Various forecasts for 2030 give a sea level higher by 4–40 cm in relation to the current one, while the equilibrium level that corresponds to the doubling of equivalent CO_2 entails probably a greater increase by 100%; this increase, though, will take many decades before it is completed.

On the hypothesis that the phenomenon will continue and that the scenarios that predict the highest temperature increases will be realized, the repercussions for humanity will be serious, given also that one-third of the human population lives in a distance of 60 kilometers from the coastal zones. Flooding of important ports and coastal areas, increase of damages due to tides and tidal waves, alteration and erosion of coasts, salinization of the water table, wetlands' degradation etc. are expected. Dealing with such hazards requires costly protection measures that—at least for developing countries—are considered to be infeasible.

The consequences for global agricultural production would also be significant, although the complexity of the phenomena does not allow safe conclusions. Agriculture is closely adapted to the existing climatic conditions and is extremely vulnerable to climate changes; every significant change would probably be detrimental. The increased presence of CO_2 in the atmosphere and the increase of rainfall might promote the growth of some agricultural plants, but also that of parasites. The changes in the water cycle will intensify the already great problems of desertification and soil erosion and will bring an extension of the water scarcity zones with adverse consequences for the agriculture of the respective areas. The increased frequency of natural disasters, such as heat waves, droughts, fires, floods, gales etc. will also be a significant adverse factor for agricultural production. However, the estimations about the results of climatic change on agriculture are still considered uncertain.

A series of other consequences are considered probable, such as forest displacement closer to the polar regions, the perturbation of many natural ecosystems, a greater risk of flooding in cities and industrial areas, degradation of life conditions (heat, humidity, discomfort) in cities with a tropical or Mediterranean climate, increased energy consumption for air-conditioning systems and the disablement or operation difficulty of many technical works (dams, reservoirs, irrigation systems etc.). Our current knowledge about the greenhouse effect does not allow specific or accurate predictions.

Part 3

Management of
Environmental Problems

Chapter 10

Pollution Control Technology

DOI: 10.1201/b14609-10

Today, the techniques for the control of various pollution forms constitute a complex advanced technology with great possibilities. Pollution control can be achieved by methods aiming either at the decrease of generated pollutants, i.e. through preventive action, or at better treatment and disposal of generated pollutants, i.e. through repressive action. Some facts in relation to the technologies for dealing with pollution caused by liquid, gas, and solid waste are discussed here.

10.1 DEALING WITH WATER POLLUTION

Various systems for treatment and disposal of wastewater are used in dealing with water pollution from the liquid waste of cities.

Liquid waste (wastewater) of cities is usually collected by a subterranean drainage network (sewerage network) and conveyed to a facility where it undergoes the proper treatment for each case. Disposal comes next, i.e. wastewater is dumped into an aquatic receptor which can be the sea, a river or current, a lake, the subsoil or underground water. Wastewater may be directly reused, e.g. for irrigation. In some developed countries, the separation of wastewater into "black" and "grey" is promoted by the use of separate networks. The black wastewater corresponds to the more burdened liquid waste, e.g. from toilets, while the grey wastewater corresponds to liquid waste, e.g. from washing machines, which is easier to reuse.

The kind and degree of treatment of the cities' liquid waste depend on the characteristics of the receptor or the kind of reuse. A minimum required treatment, as a rule, is the greater or smaller removal of suspended solids and organic matter (BOD, biochemical oxygen demand). Those removals take place in primary treatment units and can be followed by secondary treatment units.

In Figure 10.1 various treatment stages are depicted; the required degree of treatment can be achieved through a proper combination of the stages. The primary treatment stage usually includes screens, sand and grease

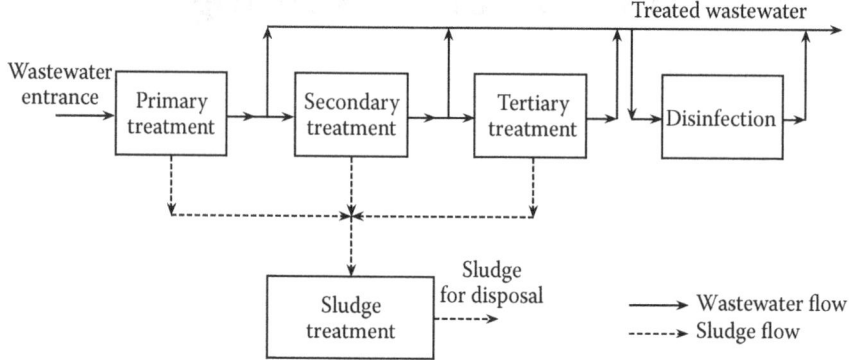

Figure 10.1 Functional model of a wastewater cleaning facility. (From Hadjibiros, K. (2007). *Ecology. Ecosystems and Environmental Protection,* 3rd edition. Symmetria, Athens (in Greek). With permission.)

removers, and primary sedimentation tanks; it is based on physicochemical processes and achieves a 50–80% suspended solids removal and a BOD removal by 30–40%.

The secondary treatment stage is mainly based on biological processes. This stage consists of the biological reactor and the final sedimentation tank.

The tertiary treatment stage is a combination of several units that varies according to the pollutants undergoing treatment. Such units are, e.g. the filters or the activated carbon facility. The removal of nutrients contained in wastewater (mainly nitrogen and phosphorus) can be carried out at this stage with the help of biological processes.

The purpose of disinfection is to destroy pathogenic microorganisms in wastewater and is applied when there is a risk of contagion due to the receiving water uses (irrigation, swimming etc.).

The sludge treatment stage is useful for the proper treatment of sludge resulting from the other wastewater cleaning stages, so that its disposal may become easy and safe. It usually consists of thickeners, digesters and the dewatering system. Sludge is a by-product of wastewater treatment that needs further processing so that its disposal will be easy and safe in agriculture, in waste to energy facilities or in brownfield area regeneration.

The primary aim of biological treatment is the removal and breakdown of organic substances through the metabolism process carried out by various microorganisms, mainly bacteria. Part of the organic substance gets oxidized in order to supply the metabolic processes with energy, while the rest of the organic matter is used for synthesis, i.e. for the creation of new microbial mass.

According to the method of oxidation, we distinguish two types of treatment: aerobic and anaerobic. In the first, disintegration is achieved by aerobic microorganisms in the presence of oxygen; it takes place at a relatively high speed and the greatest part of the organic matter is used for synthesis. In the second, disintegration takes place via the help of anaerobic microorganisms, while the greatest part of organic matter is transformed to inorganic compounds and methane. For wastewater treatment, aerobic biological processes are usually used. Anaerobic processes are generally used less often, but are used more often for sludge treatment.

The possibility for biological treatment mainly depends on the following factors:

- The degree of wastewater's ability for disintegration
- The presence of necessary nutrients, such as nitrogen, phosphorus, vitamins and various trace elements
- The pH, which should range between 6 and 9
- The absence of toxic substances; their concentrations should not exceed certain limits beyond which they are destructive for the microorganisms

In many industrial liquid wastes, as opposed to the usual urban wastewater, the above preconditions are not met. Thus special pre-treatment processes are often required for industrial liquid waste as for example for metal and other toxic substance removal or for pH neutralization. After this, those wastes can usually undergo biological treatment.

The most often used processes for biological treatment of wastewater are the activated sludge system (Figure 10.2) and the biological filter; biological disks can also be used as well as various types of stabilization ponds and artificial wetlands. The outputs achieved are high and they can reach a cleaning degree of 95%. For sludge stabilization, aerobic or anaerobic digestion is applied, the first being the most usual one.

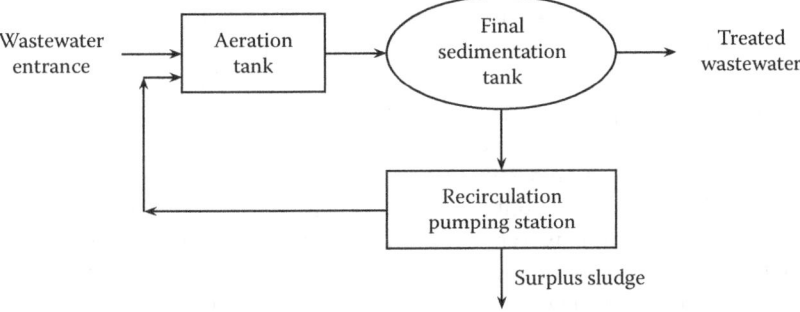

Figure 10.2 Activated sludge system. (From Hadjibiros, K. (2007). *Ecology. Ecosystems and Environmental Protection,* 3rd edition. Symmetria, Athens (in Greek). With permission.)

10.2 DEALING WITH ATMOSPHERIC POLLUTION

Atmospheric pollution at a planetary level but also at the level of cities is one of the most serious environmental problems we face today. Many technological solutions have been tried in the last few years either in the direction of cleaner methods of production or in treatment methods for various air pollutants. Considerable results have been achieved in many cases; however, air pollution, especially at a global level, is a problem that will not be solved anytime soon.

Dealing with the atmospheric pollution of cities presents increased difficulties because of the large number and variety of the sources of pollutants' emissions. The automobile usually constitutes the major pollution source in the cities. The dependence of man on that convenient means of individual transportation, the great multitude of cars and the operating conditions of the engine (that present many fluctuations and are difficult to control) render control of pollution caused by cars a difficult and complex technical, economic and social problem. Industry generates various atmospheric pollutants by many different procedures. The greatest problems are related to pollutants from burning processes. Consequently, the use of improved fuels could be a significant tool for dealing with industrial air pollution.

One measure of fuel improvement that has been applied and is effective for dealing with atmospheric pollution caused by cars is the abandonment of lead as a gasoline additive, with a gradual transition to the general use of lead-free gasoline. Another useful measure would be the proper technical modification by aiming at shortening the time of achieving conditions of normal engine function, because great quantities of hydrocarbons (unburned gasoline) and carbon monoxide are produced in the first 30 sec of an engine's function. An important measure for the decrease of SO_2 pollution, both by industry and cars, is the use of fuels—especially petroleum—with decreased sulphur content. However, CO_2 pollution will not be significantly reduced by technological measures while fossil fuels are being used.

In addition to the above improvements, radical changes in the kinds of fuels are being considered, such as the use of CH_3OH (methanol) and H_2 (hydrogen). Burning of methanol does not reduce the charging of the atmosphere with CO_2, but if methanol is produced from the CO_2 of the atmosphere (e.g. through biomass from agricultural waste), the relevant balance will be null. H_2 is a fuel that does not generate harmful pollutants, since the waste of burning is water, while the generation of NO as a by-product in hydrogen engines is negligible. H_2 can be conveyed via pipes just like natural gas for domestic or industrial use, at a relatively low cost. However, for its wide use to be promoted, its production cost from renewable energy resources should be reduced, combustion technologies must be reliable and accident risks (e.g. explosions) should be minimized. Another noteworthy possibility is electric cars, which still need improvement in the sectors of cost and performance. In general, the use of electric energy for transport as well as for other urban energy needs could drastically improve the atmosphere of cities. Of course, the production of electricity will environmentally burden some other areas, while the production of cleaner fuels requires energy consumption. Nevertheless, it is generally easier to deal with the pollution caused by a few assembled energy production stations than the pollution caused by cars and other scattered sources of burning, especially inside the city. Many specific technical and financial problems should be resolved in order for the above possibilities to become effective actions.

The other technological direction for dealing with atmospheric pollution is the removal of pollutants from flue-gases through the proper treatment, their better dispersion in the atmosphere, e.g. with tall chimneys, and the creation of conditions for more proper burnings so that CO, SO_2, NO and unburned hydrocarbons' emissions are reduced.

Industry applies a great variety of treatment methods depending on the kind of pollutants and also on the kind and the scale of the industrial unit or other special conditions. For the removal of SO_2 various methods are used: binding of sulphur in ammonium sulphate by the use of ammonia, conversion of SO_2 to H_2SO_4 under suitable temperature and humidity conditions in the presence of fine-grained coke by which it is absorbed, addition of lime in the burning chamber and then removal of the formed $CaSO_4$ dry dust through leaching etc. The removal of nitrogen oxides in industry is possible by, among other methods, their absorption by alkaline solutions and their reduction to ammonia in the presence of a catalyst. When the oxides are produced by burning, decrease of their production can be pursued by proper design and function of the burning system. The removal of suspended solids can generally be carried out by channelling the gases through a suitable treatment chamber where a force (gravity, electrostatic, centrifugal) removes them from the gases' current. CO and hydrocarbon emission control is achieved in industry by suitable conditions of burning and by complementary facilities for burning the generated exhaust fumes.

During the last decade of the 20th century, significant improvements in gasoline-powered cars were applied thanks to the use of catalytic converters made of noble metals, together with lead-free gasoline. The operation of catalytic converters is impossible if the gasoline contains lead. Three-way catalytic converters facilitate the oxidation of carbon monoxide and hydrocarbons to carbon dioxide and water vapours, while at the same time they achieve the reduction of nitrogen monoxide to molecular nitrogen gas. The result is a significant decrease of the atmospheric pollutants of cities, i.e. CO, NO_x (nitrogen oxides) and hydrocarbons and the secondary pollutants of photochemical smog. The performance of catalytic converters in ideal conditions reaches 80–90%; to achieve this performance good maintenance is necessary; when the catalyst is cold, high performance cannot be achieved in a short drive. It should not be expected that the decrease of a city's atmospheric pollutants will reach corresponding high percentages even if the technology of catalytic converters is applied globally. Their use is definitely beneficial but it is not enough to counterbalance the continuous increase of cars circulating in a city.

Consequently, in addition to technological measures, measures for the reduction of circulating cars are required. The first step in this direction is for the city to acquire attractive, i.e. fast, comfortable and safe, means of public transport. The subway, other track-guided transport vehicles (trams) and separate fast lanes for buses increase the attractiveness of mass transport and have been successfully applied in many cities. As a complementary measure, limitations to private car circulation can be enforced as well as to other polluting activities. Interventions of a town-planning character for better organization and management of the city are very effective as are the organization of pedestrian street networks, controlled parking systems etc.

10.3 URBAN SOLID WASTE MANAGEMENT

Urban solid waste (USW) management requires a complete system that includes temporary storing, collection, transportation, treatment and final disposal. The inadequacy of the existing technological solutions sets off the need for social participation in USW management.

A fictitious technological solution that could minimize social participation is represented by a hypothetical robotic mechanism. It would pick up USW left by citizens in a bin without any sorting, and it would separate them diligently and accurately into all the necessary categories so that their full valorization could be possible. However, since this ideal robot—i.e. the proper mature technology—does not exist, the active social participation for the separation of USW at source is necessary for the serious environmental burden and the negative social attitude towards USW treatment facilities and landfill sites to be avoided.

Proper design and the employment of land use policies contribute to the complete organization of USW management. When urban centres grow without appropriate planning, the siting for uses like spaces for USW sorting at source, recycling centres, final disposal facilities etc. is difficult. Also, the formation of a tight urban fabric, with insufficient arterial routes, a narrow road network, and circulation and parking difficulties, hinders the activities related to USW management. The situation is aggravated by the social reactions displayed by the residents of neighbouring areas against the siting for USW treatment or disposal facilities. Those reactions are justified to a great extent because of the serious nuisances created in an area by the continuous passage of rubbish collection vehicles, the odours, the view of landfills, the ill feeling created by the adjacency of rubbish, the concern

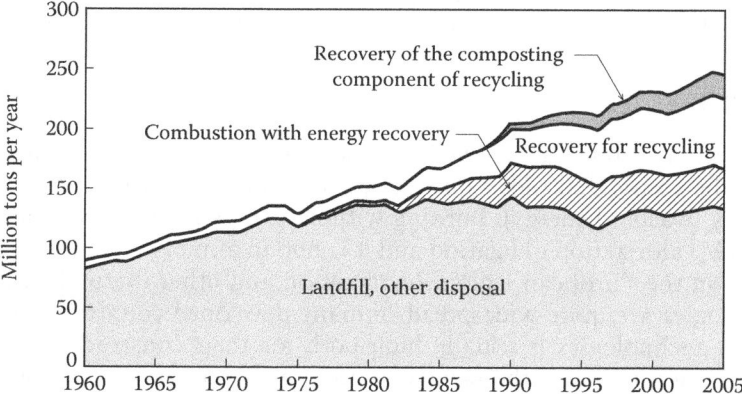

Figure 10.3 Trends in municipal solid waste generation, recovery, and disposal, from 1960 to 2005. (From Masters, G.M. and W.P. Ela (2008). *Introduction to Environmental Engineering and Science*, 3rd edition. Pearson, Prentice Hall, Upper Saddle River, NJ.)

about possible hazards or accidents etc. They are also associated with the unavoidable conflict of local interests with the greater social interest.

The possibilities for USW valorization or disposal are defined by the existing technology, the organization and the degree of social participation aiming at sorting at source. They may be (Figure 10.3):

- Uncontrolled burial
- Incineration (with or without waste to energy technology)
- Sanitary landfill
- Mechanical sorting
- Sorting at source, recycling, composting

10.3.1 Landfills

Uncontrolled disposal or unorganized burial in landfills is a traditional way of getting rid of rubbish. It has been applied for centuries by all human societies. In older times, it did not create problems since solid waste was mainly biodegradable food scraps. Even today, it is applied on a large scale in most of the settlements of developing countries and in several developed ones. Societies often overlook the impact of uncontrolled disposal because they do not realize the environmental hazards caused by modern USW composition alteration. Today even the least developed societies use modern products and so their rubbish has a very different chemical composition in relation to the past. Need and ignorance as well as the feeling that it is a temporary situation allow for great margins of social tolerance for uncontrolled landfills, contrary to the strong opposition often displayed against the institutionalization of organized landfills.

10.3.2 Incineration

Incineration or burning, as well as other thermal treatment technologies, should be considered today only on the condition that they allow energy recovery (waste to energy). Burning without energy recovery is contrary to the USW valorization obligation and it is not an aim of their management, at least in the European Union. Incineration and other thermal treatment technologies are quite widespread in many developed countries. The cost of these technologies is usually high both for their construction and for the operation phase of the facility. Many objections have been voiced for the environmental impact of USW incineration. It is true that it can cause serious problems of air pollution since a great variety of gaseous toxic waste is produced, including dioxins generated from the incineration of plastic substances that contain chlorine. Modern pollution abatement technology combined with high-temperature burning can drastically decrease the relevant air pollution. However, the use of suitable technology and systematic maintenance so that the facility may perform in a satisfactory way significantly increase its construction and operation cost. Other parameters that are important for the choice of incineration are:

- The long-term commitment to this technological choice, because the cost of the facility is very high and must be paid fully from the start
- The need for designing the facility for the maximum expected USW flow, which creates difficulties if USW production displays great fluctuations, e.g. seasonal, as usually happens in tourist areas
- The need for USW to have a significant thermal power so that there may be energy production without the need for fuel addition, which entails not only absence of recovery but also a greater operation cost

10.3.3 Sanitary Landfill

Sanitary landfill is a technology that differs from uncontrolled dumping in landfills. USW is disposed in layers that are then covered with suitable topsoil, while at the base of the sanitary landfill there is proper waterproofing material for the protection of the water table. The produced liquid waste (leachates) and gaseous waste (biogas) are collected with the proper pipe systems. It is a relatively low-budget method when there is cheap land and if the imposed burying fees are not high. It is considered suitable for regions where rubbish is characterized by a high content of humidity and a great seasonal fluctuation of their quantity and where tracts of non-productive land are available.

However, it is a technology that does not aim at the recovery of materials or energy and requires difficult and long-term maintenance. The sanitary landfill remains biochemically active for decades because of organic matter fermentations that produce toxic substances and greenhouse gases. Leachates and VOCs (volatile organic compounds) contain dozens of toxic substances that pose a threat of polluting underground water or the atmosphere. The generated biogas contains mainly carbon dioxide and methane, contributing significantly to the greenhouse effect. It is estimated that rubbish burial constitutes one of the major methane sources with a contribution of 7–20% in the global anthropogenic emissions. The sanitary landfill requires systematic monitoring and scrupulous management for several decades after the cessation of rubbish disposal and until it becomes biochemically inert. Obviously, such a long time frame carries the significant risk of failure of the waterproofing or the pipes due to poor initial construction, insufficient maintenance, accidents, or natural disasters such as earthquakes, floods, fires etc.

Sanitary landfill is a technology that bequeaths an environmental problem, even if partially under control, to at least the next two generations; therefore it appears to be radically opposed to the sustainability principle (Chapter 13). It can be characterized as an environmentally defective technology and cannot be considered an optimal available technology. Its use is in decline in European countries. Thus it is to be expected that the siting of a sanitary landfill presents a serious lack of social acceptance. It will be very difficult for people to accept having such a facility in their neighbourhood when dozens of trucks will be coming and going. It is also very difficult for the citizens who live near an idyllic gorge to accept its conversion to a new sanitary landfill. The difficulty of even the most ambitious programme of sorting at source is smaller than the huge difficulties that an endeavour for the construction of sanitary landfills entails; these endeavours often fail due to the lack of social acceptance.

10.3.4 Composting

Composting is a biostabilization method. It has to do only with the fermentable part of USW, that is, basically the residuals of foodstuffs and garden waste (trimming and pruning). Through partial decomposition, a stabilized organic material is produced, i.e. compost, that is relatively inert, does not have a disagreeable odour, is to a great extent free of pathogenic microorganisms and has a high capacity for withholding humidity. These qualities render it suitable for soil ameliorant in eroded or degraded ground and in brownfields or even for organic fertilizer in agricultural cultivations. Therefore, composting can be considered as a satisfactory valorization method of USW since it produces a useful product. An alternative possibility for composting is its small-scale implementation in the gardens of residences, with the use of special bins. This solution gives an excellent product, and it is easy and inexpensive.

Acceptance of composting products by prospective buyers, such as farmers, is problematic if the material is not of good quality. A necessary prerequisite, though, of good quality is that the fermentable part of USW should contain only food residuals and garden wastes without an admixture of other materials, e.g. plastic or glass. This purity is achieved only when the fermentable part of USW has come from careful sorting at source, in which case it becomes acceptable and can even be sold.

10.3.5 Mechanical Sorting

Mechanical sorting is a technology that aims at freeing the citizen from the obligation of sorting USW at source, while simultaneously it tries to achieve the utilization of a part of USW for energy production or as raw material. Sometimes it appears by the name *mechanical recycling* which is misleading since the products are of poor quality and therefore not useful; the production of acceptable recycling products is not possible without sorting at source. Mechanical sorting refers to a set of techniques through which USW processing is attempted so that it can be separated into metals, fermentable materials and materials for incineration. Fermentable materials are put through a biostabilization process where compost intended for soil ameliorant is produced. However, since the fermentable materials that have resulted through mechanical sorting contain admixtures, the produced compost is of poor quality and is not accepted by farmers.

Materials for burning are transformed into a product intended as a fuel substitute. Again, though, due to imperfect sorting, the fuel substitute produced is of poor quality and is generally not acceptable for energy production.

The products of mechanical sorting, both compost and fuel substitute, usually end up in burial because there are no opportunities for sale or some kind of utilization by public or private users. Even free-of-charge disposal fails because the potential user must be burdened with the cost of transportation of a low-quality product from the location of production to that of use, which leads to a minimization of the quantity that he would be interested in utilizing. Moreover, experience shows that mechanical sorting plants have a high construction cost, have frequent operation failures and their performance is generally low.

10.3.6 Reuse

Reuse applies only to some products or packaging, e.g. glass bottles, that can be used more than once in the same system. There is an indisputable environmental advantage associated with resource-saving and waste reduction. However, in evaluating the method, one should take under consideration the potential environmental burdening from the cleaning, the disinfection and the transportation of reusable objects and also the fact that reusable packaging is usually sturdier and heavier; thus it requires greater consumption of energy and other natural resources.

10.3.7 Recycling with Sorting at Source

Recycling with sorting at source is the friendliest to the environment management method of USW. It diminishes environmental burden and contributes to natural resource saving. Urban solid waste suitable for recycling is mainly paper, glass, metals, plastics and construction and demolition waste. The citizen makes sure to dispose of the various materials in different bins. In parallel, he can also apply a system of domestic or municipal composting of fermentable waste. An additional important advantage is that sorting at source turns the greatest part of USW from rubbish to raw materials and in this way the negative environmental impact from the transportation of USW by collection vehicles is minimized.

Sorted-out materials are received by corresponding industries that produce recycled paper, metals, glass etc. Recycling with sorting at source is technologically and financially feasible, although for certain materials, as for instance glass, its cost effectiveness is marginal and sometimes the need for subsidy may arise. Cost effectiveness for glass, paper etc. improves if further sorting into different qualities is carried out.

Recycling is a widely applied and rising technology in most developed countries, where it is promoted with priority. It is compatible with the increasing environmental protection requirements as well as the evolving European environmental policy. A recycling programme with sorting at source should include special actions for the change of social behaviour and achievement of social participation, effective organization of collection and disposal of recyclable materials and provision for management and gradual decrease of the waste quantity that will be created by imperfect sorting. The cost and materialization time of such a programme are definitely smaller in comparison to other costly and ineffective solutions such as mechanical sorting or sanitary landfills that often cannot achieve social acceptance.

10.3.8 Recycling by Manual Sorting

Another form of USW treatment is recycling with partial sorting at source completed by manual sorting in sorting centres. The citizen takes care to dispose of all recyclable materials together, e.g. packaging, in special bins from which they are picked up by special vehicles and taken to the sorting centre. There, the centre's personnel performs manual sorting of the materials on a conveyor belt. After that, the materials are picked up by the corresponding industries for recycling. This method cannot succeed without an initial partial sorting at source. Manual sorting cannot be applied to mixed wastes, that is, when they include a mixture of fermentable, recyclable and other materials; it is also impossible to be applied to waste that has been compressed in transfer stations.

10.3.9 Residuals Landfill

A complement of recycling is residuals landfill where what remains of USW is carried for burial after the recyclable and organic fermentable substances have been removed. If this residual (5–20%) is biochemically inert, its burial does not entail generation of significant quantities of liquid or gaseous pollutants. Therefore residuals landfill has a small environmental impact and can be socially accepted.

10.3.10 Special Solid Waste Management

Recycling activities, in a wider sense, include management with material recovery from special solid waste, such as batteries, electric appliances, old cars, mineral oils, rubbers, catalytic converters etc. Their management is mandatory according to recent European legislation. Special management is also required for hospital solid waste as well as the sludge from biological treatment of wastewater.

10.3.11 Combination Possibilities

The various USW management methods can be combined only in certain ways, while some combinations are not possible.

- Paper and plastics sorting at source and recycling diminish to a great extent the burning possibilities of the remaining rubbish.
- USW grinding and compression in transfer stations exclude the possibility for subsequent manual sorting and recycling.
- Sorting at source followed by recycling with maximum materials recovery, in combination with composting of the fermentable USW, entails the burial or the incineration of only a small quantity of residuals (first priority scheme of European environmental policy).
- Sorting at source and recycling of certain materials (e.g. metals and glass) facilitate the incineration of the remaining rubbish (second priority scheme of European environmental policy).
- Application of fermentable USW domestic composting has a positive result both for the option of maximum materials recovery and for the option of incineration.

- Not sorting at source entails that incineration of the whole of USW will be more expensive and with a greater environmental burdening, composting after mechanical sorting will yield a useless product and the recovery of materials after mechanical sorting will not be satisfactory.

10.3.12 European Policy and Legislation

European policy for USW management is brought into effect by legal texts and programmes that promote the following:

- The recovery of materials and energy, with special emphasis on recycling of materials and with specific quantitative goals that increase in the course of time
- Volume reduction and maximum packaging valorization, with a corresponding minimization of burial
- Obligatory USW processing before they are taken to burial
- Gradual reduction of biodegradable USW (fermentable materials and paper) that are taken to burial, so that in the long run only an inert part of rubbish will be allowed to be carried to landfills
- Prevention of health and safety problems for those working in USW management through encouragement of rubbish separation at source and avoidance of manual mixed USW sorting

Chapter 11

Tools for Environmental Analysis, Evaluation and Management

DOI: 10.1201/b14609-11

11.1 ENVIRONMENTAL MANAGEMENT

Environmental management aims at rationally dealing with man's interaction with the environment—mainly with the decrease of the environmental impact that human activity entails. Its goal is not simply the protection and preservation of the environment but the optimal utilization of its structure and functioning for the benefit of humanity. Environmental management presupposes the recognition of the dynamic function of ecosystems; it includes the organization of regulations so that the preservation of their important characteristics may be achieved and also the modification of their structure and function may be under scientific and social control. The regulations pertain to the physical, chemical, biological, geological, aesthetic etc. components of the environment, both biotic and abiotic, as well as their relationships with the social, cultural and economic environment of man. Many different environmental management tools are available.

11.2 DEFINITION AND ASSESSMENT OF IMPACTS ON THE ENVIRONMENT

Many human activities including industrial production, technical works and agriculture degrade the atmosphere, the soil, the waters, trophic chains, biological diversity and the structure and function of the biosphere in general; they also have a negative influence on human health, cultural monuments, social conditions etc. Thus they lead to various consequences on the quality of human life, on natural ecosystems and the landscape, on the cultural, social and economic conditions etc. Those consequences are generally called *environmental impacts* and are due not only to pollution but to other perturbations of environmental systems as well. The impacts, which are usually negative but can sometimes also be positive, may be local, regional or planetary, direct or indirect, short term or long term, reversible or irreversible.

The distinction of environmental impacts into reversible or irreversible is very significant. The first ones correspond to damages that can be restored by nature or human technology to a satisfactory degree and in a time scale of a few decades, maximum; the others, however, (e.g. decrease of biodiversity, soil erosion or desertification) constitute irreversible environmental degradation, which greatly increases their severity.

Objectively examined, environmental problems whose impacts extend to the whole planet, such as the greenhouse effect or ozone layer depletion, are more important than regional ones that are manifested in large areas (e.g. acid rain) or local ones that pertain to specific cities, industrial areas, marine bays etc. Environmental problems that negatively affect a larger number of people are also more important.

Dealing with environmental problems requires an appropriate environmental policy that interacts with policies applied in other sectors, like agriculture, industry, transport, energy, tourism and, in general, all economic activities.

11.3 ENVIRONMENTAL INDICES AND INDICATORS

Assessment of the environmental state and its impacts can be performed on the basis of data from measurement networks, observations, statistics or inventories. A modern direct or indirect way of evaluation is, correspondingly, the use of proper indices or indicators that offer useful information about the state of the environment; thus, time and effort are both saved. They constitute representative assessments of parameters or values resulting from parameters, which describe the environmental state and its impacts on man, ecosystems and materials, or refer to the pressures exerted on the environment, to the causes that bring them about or to society's responses in relation to the negative impacts. They are generated by an information selection and collection process. They "condense" the information that comes from various distinct and periodic measurements or assessments thus forming a representative numerical value or qualitative evaluation. Indices can be the average or highest values of pollutants' concentrations or numerical results of calculations or other kinds of statistical results that provide direct information about the environmental state. Indicators are used mainly to carry out indirect assessments, without analysing the complexity of the phenomena; warnings about oncoming dangers are given and the measures that need to be taken are evaluated. The more and better indirect information indicators provide, the better they are. Indicators of atmospheric pollution correspond to air quality problems as well as to acid rain, toxicity, influence on the ozone layer or the greenhouse effect etc.

Less classic indicators are those that express the environmental state in a more indirect way, for example:

- The presence (Chapter 4) of significant predator populations (e.g. wolves, eagles, dolphins) in a natural ecosystem, since they provide evidence for the good functioning of the whole ecosystem
- The percentage of mothers or fathers and small children strolling through a city, because it provides evidence for the quality of the urban environment

Before they are selected for use, environmental indicators or indices should be checked by experts in relation to their suitability. The selection criteria are many and various, e.g. how significant is the specific environmental problem and how does society deal with it, how feasible is the measurement or assessment of the specific index. Therefore, indices or indicators, in order to meet the criteria as much as possible, should be:

- Representative of the problems
- Significant
- Persuasive
- Accurate
- Transparent

11.4 LIFE CYCLE ANALYSIS

Life cycle analysis (or life cycle assessment, LCA; Figure 11.1) includes a series of methods for the detection, assessment and wherever possible, quantification of environmental consequences of a product, a work, a process or activity for the duration of its life. For example, in the case of an electric appliance, the environmental aspects of extraction or production of raw material in general will be considered as well as aspects of the manufacturing processes, distribution, energy and other resource consumption by its use, of waste generation, of its potential reuse or recycling and final valorization or disposal of the useless appliance. This method can also be applied to a technical work, a manner of production, a material etc.

The two basic phases of life cycle analysis are identification of the important components and assessment of their impacts. Identification of important environmental components aims at the quantification of environmental inflows and outflows during the product's life span. The result is a list of pollutants or other burdens that can cause a negative impact on the environment. Usually, those burdens are not comparable to one another. Impact

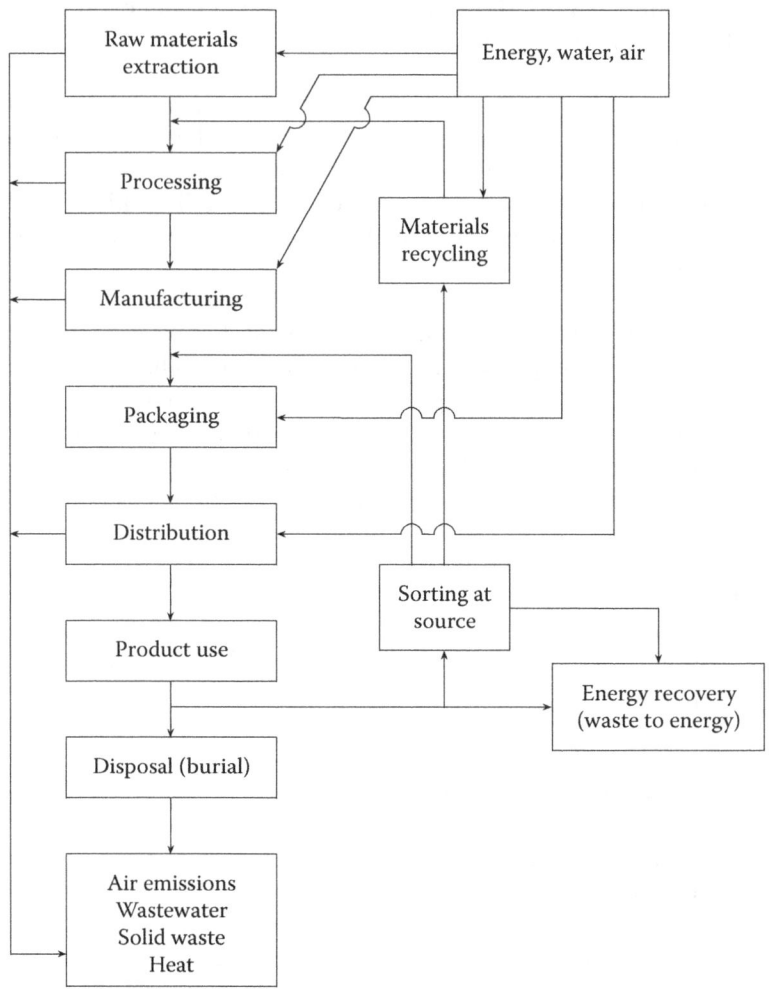

Figure 11.1 Life cycle of a product.

assessment aims at grouping and evaluation of the potential environmental burdens that have been identified as, for example, climatic change or eutrophication. The impacts are grouped on the basis of equivalence factors, e.g. potential for global warming or for ozone layer depletion or for eutrophication in a lake. Usually inclusion of different impacts is not obvious; in that case, corresponding values of a quantitative or qualitative character should be set according to their relative significance. The various evaluation methodologies used can be linked to goals or costs or be assessed by a group of experts or people involved with the subject. For the time being, there is a lack of commonly accepted evaluation methodologies.

11.5 ENVIRONMENTAL AUDITS

An environmental audit is the examination of the degree of compatibility between the manner of management and operation of a business or organization with the environmental protection requirements, the generally proper environmental practices or the sustainability principle. In some cases, environmental audits are mandatory by legislation. They are applied as a tool for exploring the effectiveness of the environmental efforts made by businesses or organizations at a local level.

These systematic, independent examinations aim at finding to what degree environmental outcomes correspond to the goals, or how effective the methods for achievement of the goals are. Environmental audit is a study of records and documents as well as a discussion with people who hold key positions in the organization, so that possible deviations between goals and outcomes may be identified.

11.6 ECO-LABELS

Eco-labelling is a promotion method for environmentally friendly products and services. The eco-label is useful for distinguishing a product or service that is preferable among a variety of similar products or services because it possesses some positive ecological characteristics. The distinction is based on examination of the life cycle of the product or service and is granted by an independent body on the basis of ecological criteria. There are many different labels, both optional and mandatory, that confirm ecological performance. Eco-labels serve a common goal—the encouragement of supply of and demand for products and services that cause less pressure to the environment, through reporting verifiable, accurate and not misleading information in relation to their environmental aspects.

The roots of eco-labelling can be found in the increasing concern of governments and public opinion about the environment. Businesses recognize that environmental care can bring advantages to marketing of certain products and services. Various eco-labels have appeared in the market, such as natural, organic, recyclable, friendly to the environment etc. However, reliability and objectivity are needed, and these can be guaranteed only by the intervention of certified public or private organizations that bestow the eco-labels.

11.7 ENVIRONMENTAL CERTIFICATION

An environmental management system aims at reducing environmental impacts that are assessed by certain objective criteria, i.e. to manage the environmental risk coming from an activity. The formation of an environmental management system for a business, organization, a governmental agency etc. can be based on certification. Certification systems offer guidance and advice to a wide spectrum of sectors that include audits, labelling, life-cycle assessment etc. In the last decade, the ISO 14001 system has been used at an international level for environmental certification. It is part of the ISO (International Standards Organization) series. At a European level, EMAS (Eco-Management and Audit Scheme) has been developed on the basis of a European Union resolution that requires its application in all member-states. For the time being, it is an optional system aiming at certifying businesses, organizations etc. It provides recognition that the organization has established an environmental actions programme designed for protection of the environment against the negative impacts of the organization's activities and for continuously improving its environmental performances.

Resource squandering and polluting activities in some sectors of a business are evidence that it needs significant environmental improvement. All systems of environmental management focus on the need for continuous and unlimited effort for minimization of environmental burdening for the sake of the quality of life of present and future generations. A connection with a system of environmental management pertains to all categories of people involved with the business's activities, specifically investors, employees, customers, insurers, auditing mechanisms and non-governmental organizations.

11.8 GEOGRAPHIC INFORMATION SYSTEM

An environmental Geographic Information System (GIS) describes a locality by using different information levels that it combines in order to offer a better understanding of environmental problems. It is a useful tool that can assist decision-making procedures and the formation of an environmental management strategy. The selection of the information levels that will be used depends on the object of study, e.g. depiction of a point-source pollution event in a stream, designing of a regional zone of fauna protection area, exploration of a possible relationship between environmental factors and an alteration of the morbidity index of the people of an area. The spectrum of possible GIS applications is wide and covers most of the environmental sectors.

The GIS offers possibilities for better natural resource management. It can combine various groups of environmental data with the data of the locality, such as addresses or geographic coordinates, which gives the public, private companies, non-governmental organizations and governmental agencies access to the information. Through a distributed database, the information can be collected once and used many times.

11.9 ENVIRONMENTAL IMPACT ASSESSMENT

Various technical works modify the environment in order to achieve goals that are desirable by human society. However, due to certain sensitive environmental characteristics as well as to the imperfection of plans and technologies, undesirable environmental changes can result as side effects of human interventions.

Environmental impact assessment (EIA) is the compatibility examination of various developmental plans, programmes, works or other activities with the environmental protection requirements. It is carried out in the philosophical framework of the sustainable development principles (Chapter 13) with the purpose of finding the optimal solution for each case that is usually a rational compromise. The EIA is a lengthy procedure that includes one or more studies, public hearings and social dialogue procedures, definition of environmental terms, and application and monitoring of remedial measures.

Avoidance of unpleasant or destructive impacts on the environment by various works or activities can be facilitated if a proper study to predict and evaluate these activities is carried out in advance. Environmental impact studies or statements (EIS), especially at a local level, have been an important tool for the protection of the environment during the past several decades, mainly in developed countries. All important works and activities should be examined during the design phase in relation to potential environmental impacts, and hence can be modified in time so those impacts can be minimised. If the impacts are serious and their treatment questionable, the advisability of the work should be re-evaluated and the possibility of its cancellation examined. The effort for the assessment and decrease of environmental impacts in advance is important because it contributes to prevention of environmental damage, which is generally easier and cheaper than dealing with them ex post facto.

The process of evaluating and dealing with the environmental impacts of a work or activity does not stop at working out an EIS. The necessary examination of its content by competent agencies is followed by a public hearing procedure, that is, its presentation to the concerned citizens and the discussion of their questions or objections. Public hearing of the EIS is mandatory on the basis of European legislation. Next, the application of the protection measures that have been designated by the EIS should be systematically monitored.

The environmental impacts of a work or activity are usually related to the function of ecosystems and pollution. They should also relate to aesthetic, cultural, social or economic issues. The proper elaboration of an EIS is an interdisciplinary work that has to be based on several specialties and approaches, to achieve their synthesis and propose suggestions for the protection of the environment, co-assessing correctly the various factors.

The basic objective of an EIS is not so much the description and analysis of the existing environment as the prediction of its course in the future under the influence of the work or the activity. The difficulty of prediction is due to many factors including insufficient data, complexity of the environment etc. A complete assessment also requires examination of the environmental impacts of all the alternative solutions for the project because it is possible that some works may be less harmful to the environment even if they are found wanting financially or technically.

City issues (Chapter 6) that need EIS are the siting of wastewater treatment plants and waste disposal sites and the siting of large facilities as, for example, hospitals, roads or other infrastructure construction etc. Many of those works cause strong reactions from the affected inhabitants and the relevant EISs have to face difficult problems that sometimes turn out to be unsolvable. The only radical solution for such difficulties is their prediction as early as possible through proper initial design, which is usually a siting or town planning study. Any city expansion study or other town-planning regulation, such as the increment of building height, built-surface ratio or site coverage ratio, should not be conducted without an environmental impact study that in this case will have to be strategic in nature.

11.9.1 A Hydraulic Work Impact Example

An engineer is often called to carry out works of water resources management, some of which have direct, significant and usually non-reversible environmental impacts. Hydroelectric, irrigation, drainage, torrent regulation and river diversion works are characteristic examples.

Water is a vital element for the whole biosphere and especially for terrestrial ecosystems, for which it is usually a major limiting factor. Its ceaseless flow causes significant actions both biological and mechanical or physicochemical. Storing winter water discharges and flood waters is often necessary because of the shortage of water in many areas. Paucity of water resources would be less serious if the observed water squandering, mainly in intensive agriculture and to a smaller degree in industry and at home, was avoided. Any intervention attempt in the water flow can have complicated and unexpected consequences, direct or indirect, local or generalized, short or long term. Interventions could cause significant ecological or economic damages, to the point that their advisability or viability may be questionable. Thus it is necessary for a well-documented EIS to be carried out in advance, which will explore all the predictable consequences of the water management project and assess the economic and environmental risks. An example of the issues that a study like this should examine is mentioned below.

Let us suppose that the construction of a large hydroelectric or irrigation work is being planned. Its main component is the construction of a dam in a river bed with the purpose of creating an artificial lake that will serve as a reservoir for the production of hydroelectric energy, for irrigation or water supply. Throughout the world, thousands of large dams have been constructed (e.g. the Aswan dam in Egypt, the Itaipu dam in South America). The following list is not complete but is indicative of the environmental impacts' variety and complexity of such a work:

- The artificial lake will inundate areas that may include settlements, agricultural cultivations, other economic activities, historical monuments, remarkable landscapes, biotopes etc. If salvaging some of them is deemed necessary, then they will have to be transferred somewhere else if possible, or the extent of the lake should be reduced by suitable auxiliary dams.
- The dam will cause retention of river sediments and their accumulation in the lake. This entails that the whole work will have a limited life span. If the assessments about the inflow rate of sediments are misjudged, there is a risk that the lake will be filled faster and that the project will be prematurely rendered useless.
- Retention of sediments in the lake means that they will not reach their normal destination, that is, the flat riparian areas and the delta formed at the river mouth. Consequently, there will be a decrease of nutrients in the riparian and marine areas of the river mouth, possibly resulting in reduction of agricultural and piscatorial production. The reduction of sediments also poses the risk of affecting the evolution of the coastline and causing increased erosion of the banks downstream of the dam.
- Creation of an artificial lake may cause alterations of the local climate, with a possible humidity increase and temperature fluctuation decrease.
- In the area of the artificial lake, erosion and torrential phenomena will be positively or negatively influenced.
- The river flow downstream of the dam will be artificially determined, which can have negative consequences for the downstream ecosystems. This danger is increased during the filling of the lake.
- The lake water will have reduced oxygenation in relation to the river water; therefore, it will have reduced pollution assimilation capacity, especially if there is also lake stratification. There will be an increased possibility for eutrophication. If the river upstream of the dam is a receptor of pollution (municipal waste, livestock farming wastewater, fertilizers etc.) there may be a need for more advanced wastewater treatment due to a smaller self-purification capacity of the lake water.

- The aquifer level, the quantity and quality of underground water in the area, will be altered. Some springs may go dry or new ones may be created.
- The lake will hinder the traffic of terrestrial fauna. The dam will hinder the traffic of the river fauna, which can be fatal for some populations of anadromous fish. This consequence can be partially avoided if an appropriate parallel channel is constructed that will bypass the dam.
- The change of the environment due to the creation of the lake may promote the propagation of biological species that are foreign to the area, especially parasites, harmful insects, microbes etc. The use of biocides for fighting undesirable invaders can decrease the quality of the water.
- The dam construction work will cause significant perturbation to the local ecosystems, mainly to the wild fauna, through excavation and construction site activities. It will also cause degradation of the natural landscape, especially if there are borrow areas and debris disposal.
- According to some views, accumulation of a great mass of water such as a lake can cause local earthquakes due to alterations of the underlying geological strata.
- Artificial lakes have positive environmental aspects as well. Many people believe that a created landscape has greater aesthetic value than the original natural area. Some artificial lakes turn into remarkable natural biotopes over time.

11.10 INFORMATION ACCESS AND PARTICIPATION IN DECISION MAKING

Public hearings are an important part of an environmental impact assessment process. Access to information, the ability to appeal to environmental courts and participation in decision making are effective social tools for the rational environmental management aiming at a better quality of life. In 1998, the Aarhus Convention was signed by many European countries; the Convention specifies the public's right to have access to information in relation to the environment and to participate in decisions that affect the environment.

At one time, information about the environment and the environmental impacts of various activities was kept secret. According to the Convention, this is no longer permitted and information is free, with the exception of some special cases. Availability of information is vitally important, as is its collection and publication in an easily understood and accessible form to the public. The most efficient ways are through databases and the Internet.

Citizens must be able to appeal to environmental courts claiming a fair judgement when, accidentally or intentionally, their right to the environment is violated. The Convention provides the right for an appeal against refusal to supply environmmental information by the relevant parties, against an insufficient participative procedure or against acts that violate environmental laws.

The participation right pertains to decisions about major works, such as big roads, power stations, dams and industrial facilities, as well as plans, projects, policies and legislation that affect the environment. The Convention defines the minimum level of participation opportunities necessary, and the procedures that have to be maintained.

The public's involvement is necessary for the promotion of sustainable development policies. When involvement takes place in an early enough stage, it contributes to consensus building because the opportunity is given for opposite views to be heard and taken into consideration. Utilization of the public's knowledge, skills and enthusiasm is useful for better decision making, that is, decisions that meet the citizens' needs and are more reliable and applicable for a longer time. It is well known that in some countries, the most important cause for the delay or cancelling of developmental projects is social reaction. The most effective way of dealing with this is through participative procedures.

Public is defined as one or more concerned natural persons as well as legal entities such as groups, unions or NGOs (non-governmental organizations) in general. When the participative procedure is conducted in many phases, the participation through representative legal entities that have the necessary infrastructure can be more effective.

Relationships of trust with the public are difficult to create, especially in areas where participation in decision making is a relatively new procedure. The strategy for development of the participative procedure may need a phase of trust building with the public, NGOs and the business community.

The effort for development of participative procedures should be sincere in relation to the actual possibility for influencing decisions; otherwise, the public's trust is shaken and interest for participation in future procedures is lost. The public needs to be informed about the procedure and the role it will be called to play in decision making. The public does not have a uniform view; there may be important differentiations among the views of the citizens as well as between the public and the decision makers that should be heard. The participative procedure should be open so that anyone who can be influenced by the decision, or just interested in general, can participate. By definition, it is accepted that NGOs have such an interest.

The procedures for public participation presuppose that:

- There are alternative solutions (consequently, participation can influence the direction of adopting one of them)
- What is asked from the public is clear
- There is willingness for the public's views to be heard and taken into consideration in the decision-making process

It is useful for those responsible for decision making and the ones who carry out the studies to be trained in relation to the way of approaching the public and answering questions and comments. They also need training in relation to dealing with special problems, such as, e.g. aggressive behaviour of some participants or apathy or non-active participation of some others.

The project should be explained in such a way as to be easily understood by lay people. The best way is the use of examples from everyday life. The information presented should be of good quality, complete, easily understood and accessible to the public. It is important for the public to have a good comprehension of the proposal. This goal is served by a non-technical summary. It is also important to use many speakers and presentation techniques, e.g. colour images and figures.

The participative procedure:

- Should allow the possibility to present alternative solutions or changes to the plans, and offer terms for the realization of these changes
- Does not always end up in general agreement about what should be done and does not exclude the possibility that the authorities might make an opposing or unpopular decision
- May give rise to discussions and disputes that are not related to the object of deliberation
- Usually causes delays in the advancement of works and projects, but may avert longer delays, such as litigation, and it may offer important long-term benefits

The participative procedure should not end with the decision making. It is useful to maintain a dialogue with the public even after accomplishment of the work. Feedback is needed in order to show to the public that its comments were taken into consideration. If the final decision is different from the local inhabitants' preferences, it is important for them to understand why it was made, e.g. because a larger social interest prevailed.

Chapter 12

Biodiversity and Landscape Protection

DOI: 10.1201/b14609-12

12.1 MULTIPLE DIMENSIONS OF NATURE PROTECTION

The protection of nature from anthropogenic pressures is a particularly difficult problem because it reflects the multiple dimensions of the interaction between natural and social systems. Natural ecosystems and the landscape as well as all kinds of human activities that disturb them present a great complexity. The damage caused by pollution is a significant part of the problem (Chapters 7, 8, and 9) that is described quantitatively or qualitatively by the relevant sciences (chemistry, toxicology etc.) and can be technologically dealt with to an increasingly satisfactory degree (Chapters 10, 11). On the other hand, various alterations of the natural environment (Chapter 5) besides pollution, in addition to their complex physical, chemical, and biological, quantitative or qualitative aspects, present serious social and cultural dimensions that often compose the basic core of the problem. The protection of biodiversity and the natural landscape encounters social difficulties and needs social approaches. The related issue of cultural environment conservation can be similarly considered, since the reasons for protection and the methods employed for it are generally the same. The landscape, which is the projection of our natural environment, is also the image of our culture; moreover, the monuments, historical sites and traditional settlements give additional remarkable and unique elements to this natural landscape.

12.2 PROTECTION OF BIODIVERSITY AND THE LANDSCAPE AS A SOCIAL ISSUE

The efforts for biodiversity and landscape preservation imply limitations to human activities and therefore often are in conflict with the value of freedom. The principle that one's freedom stops where the freedom of the other begins is called upon to undergo one more limitation for the sake of life in general or for the sake of ecosystems, of natural or cultural

215

monuments etc. The need for sensible environmental management adds an extra layer of organization to the already too-organized life of people in the modern world, some of whom are suffocating under the burden of rules.

Sometimes activities for the preservation of biodiversity and the landscape are considered by those affected as an insult against the value of equality or social justice—at least as it is interpreted by them. For example, in order for an intellectual person to enjoy an archaeological site or for a nature lover to admire a rare bird in a wetland, the local farmer should limit his possibilities of increasing his income, e.g. through irrigation, drainage or building works. Similar protests are expressed in some European countries against the European Union policy, because by strict environmental protection legislation it imposes limitations without compensating those affected for the resulting loss of income.

As a rule, the development of activities that change the environment meets the public's expectations for more security and comfort in life. Security is a fundamental value for individuals and for society as a whole. The limitation of wild nature, of non-friendly biological organisms and of many natural processes has been at the centre of the endeavour for increased security in the past. In spite of the current rise of the economic and technological level and the drastic reduction of dangers that come from nature, a certain degree of opposition still exists. Comfort is also a basic pursuit of man, who tries to live better, labouring less. Thus few people, for example, would sacrifice the sense of security offered by the night illumination of a park in order to protect nocturnal animals from the disturbance caused by the lights. Similarly, few would want to put up with mosquito invasions at nightfall in order not to pursue drainage of a swamp next to their house.

Thus by seeking greater security and comfort, by demanding equality and by being aware of his freedom, man has been destroying the natural environment as much as the abilities of his technology have allowed him in the course of history. One of the first things primitive man did as soon as he developed his first tools was to clear the neighbouring forest. However, civilisation poses barriers that are associated with the pursuit of maximum collective social safety that depends also on environmental equilibrium. Those barriers limit or suspend the potential destructive inclination of the private interests of individuals. Today, the main factors that make environmental conservation practicable are the relative saturation of consumer needs in industrialized countries, the wide recognition of environmental degradation and the development of environmental concern, and the accelerating development of technological possibilities and business activities in environmental protection sectors—so-called green capitalism. Those factors are mainly manifested in countries with important economic and cultural development.

The problem of decisions in relation to environmental or other social issues is solved by the democratic principle of the majority. This principle should be applied after a dialogue that allows the minority an opportunity to argue its concerns, and problem solving should be based on the most complete possible cognitive basis. Any positive or negative decision about environmental preservation should be revised if more recent scientific data raise significant doubts about its correctness. However, the answer to the question of which body decides is not apparent. The relevant spatial (only people living in the area can decide) and temporal (only people living today can decide) limitations are not self-evident. Certainly, the local population should have the right to the first say. There are, though, many instances where a biotope, a landscape or a monument can be considered of national or even international importance, but its protection may hinder economic activities of local inhabitants. In such cases it seems that the body making the decision should be expanded to represent the national or international society. In addition, the elements of the particular environment that are of value cannot be considered property of this generation only. On the contrary, environmental values constitute a legacy that has been handed over by previous generations and will be handed down to future ones. Do ancestors and descendants have a say in the preservation of that legacy, and if they do, how should they express it? A positive confrontation of this side of the issue could be based on modern internal regulations; appropriate environmental terms could substitute older rules of a religious type which were used to preserve environmental values in the past.

On the other hand, if we accept that nature belongs to nature, then conservation becomes man's obligation in the framework of a non-anthropocentric worldview. Such a view necessitates a general respect for every living entity, probably even for every abiotic natural element as well. However, the full development of the rational consequences of this notion leads to a rejection of human progress and the acceptance of a human society of hunters-gatherers only that would be fully incorporated in the natural ecosystems.

12.3 METHODS OF BIODIVERSITY CONSERVATION

Biodiversity conservation means species conservation complemented by conservation of other ecological elements such as sub-species, genetic variety, ecosystem variety etc. This goal will be served if there is a cessation of species extinction due to human influence, that is, with reduction of the extinction rate to its natural extent. On a biosphere level this goal is not feasible for the time being, mainly because of the socioeconomic state of developing countries. It has already been achieved, though, in some developed countries where biodiversity has stabilized in a human time scale, after having been decreased in the past. The goal is realistic in the medium term in many developed countries that retain a significant biodiversity. More than anything else, its accomplishment requires that the conservation of local endemic endangered species as well as of migratory endangered species must be given a high priority. The need for research is also relevant, so that those species may be identified, the reasons that put them at risk may be defined and the proper tools to protect them effectively may be chosen.

The presence of biodiversity has an unequal geographic distribution and is strongly affected by the development of human activities. Accordingly, its protection is attempted mainly where it does not have a great affect on economic activity. One of the basic measures is designation of protected areas, that is, tracts where human activities are subjected to restrictions that are necessary for the conservation of nature.

In some views, life's network in the planet forms an indivisible and continuous whole in time and space. According to them, the attempt to characterise one part of nature as more significant than another is useless in the long term. Man's failure to maintain a balance between his activities, civilization and nature and to manage space in an integrated way forces him to take hasty measures and to isolate certain areas for special treatment versus the abandoned whole. If man was able to experience life's network as a whole, there would not be a need to preserve only a small part of the natural wealth. The only acceptable pursuit, according to this view, would be restoration of harmony between man and nature. However, only a small number of people who care about environmental conservation support such an unrealistic view (Chapter 13).

Effective biodiversity protection would require integrated management of the entire biosphere. Such an approach would consider the whole planet as a boundless protected area. Because of the great differences in biodiversity distribution, its various parts would need well organized global and local schemes of protection. If we take, for example, a country that represents the average in terms of ecological values, it could be divided into three categories of protection zones as follows:

- About 5% of terrestrial surface would need strong protection as a reserve of significant ecological values. Such areas would mainly be threatened biotopes and landscapes.
- About 30% of terrestrial surface would need careful management where mainly soft activities would prevail. Such areas would be most of the forests, several agricultural and tourist areas, many special landscapes such as rivers, alpine regions, small islands etc. Many marine areas would have a corresponding type of management.
- The remaining 65% of the terrestrial surface would constitute a zone of more intensive but controlled activities, and the protection would be based on the organization of environmental planning and the systematic prevention of degradation through environmental

impacts' assessment of works and activities, the imposition of proper terms and measures, and other environmental management tools (Chapter 11).

Such an organization of protection could ensure the cultural environment's conservation as well.

Chapter 13

Environment and Development

DOI: 10.1201/b14609-13

13.1 IMPACTS OF TECHNOLOGY

Since the era of the hunter-collector society, man's technological inventions have been changing his life drastically and have usually caused some amount of environmental degradation. After the invention of the steam engine, technology developed at a rapid rate that has been accelerating since then. Humans have acquired the power to change the environment and to improve their standard of life. In technologically advanced countries, the problems of nutrition, housing etc. have been dealt with to a great extent; man is effectively protected against extreme temperatures, floods, earthquakes and other adverse natural phenomena. Water treatment technology has almost eliminated typhus, cholera and other infectious disease epidemics that were once a scourge for human communities; vaccines have likewise eliminated many serious illnesses. Working hours have been reduced and the resulting increased free time and modern means of communication have opened possibilities to the human population for understanding the world, cultivation of the mind, travelling and communication. Abundance in developed countries has blunted previous economic and social inequalities in relation to access to consumer goods.

However, technological development has adverse side effects as well, which have become more widely recognized. For example, agricultural performance has increased through the use of chemical fertilizers and pesticides, but many surface and underground waters as well as trophic chains became polluted. Industries produce a variety of foodstuffs and other products for comfort and recreation, but many of the by-products of these items pollute the atmosphere, water bodies and the ground, with consequences that affect human health, the function of ecosystems, the climate and the aesthetic value of the environment. Cars and airplanes offer great possibilities for fast movement, and heating and cooling facilities make living and work more comfortable, but their combustion directly or indirectly pollute the atmosphere. The contribution of electric power to the quality of life is important, but power stations, both thermoelectric and nuclear, have serious consequences and pose potential hazards for health and the wider environment.

Technology has a great ability to reduce the emissions of these pollutants, but some other environmental problems, as, for example, the complex degradation that ends up in desertification or in the loss of biodiversity, cannot effectively be dealt with through technological means. Technological abilities expand through the advancement of science but they remain finite. If, for example, waste disposal from factories into a river increases, then the pollution abatement technology should greatly increase its performance in order to keep the water pollution to its initial levels. The cost of pollution abatement is high and is increasing rapidly with the increase of required performance. There is for instance a technology for the removal of sulphur from the exhaust gases of coal-fired thermoelectric power stations, but the surcharge per kilowatt-hour is high, especially when pollution abatement is applied to existing older stations. There is a technology for electricity production by solar radiation and other renewable sources, but the production and storing cost is still high. Nevertheless, future cleaner production technologies together with more efficient pollution abatement technologies are expected to create prospects of economic activities with lessening environmental burdening that will gradually tend towards zero.

13.2 THE 'DEVELOPMENT VERSUS ENVIRONMENT' DILEMMA

All human societies pursue security and comfort, and many societies or social groups demand environmental conservation, with an emphasis on quantitative or qualitative elements, in conjunction with the level of economic and social development of each society. Developed people express increased environmental concerns today, after having heavily exploited natural resources for centuries. A great number of developing countries have resisted environmental regulations for decades, asserting the so-called "right to pollution." The antagonistic relations between the environment and development can be pointed out in the framework of the rich North or the poor South, the slums or the desertified countryside, the aristocratic suburbs or the degraded quarters ("Fourth World") in big cities. The philosophical question of whether the environment should be protected as a value in itself or as an exploitable good remains open. In the first case, a qualitative approach is sufficient, whereas in the second a quantification of environmental problems is necessary. Finally, environmental conservation cannot be successful if it ignores the natural tendency of the environment for change, which is manifested in different time scales: tens of millions of years (geological tectonic movements, biological evolution) and centuries (ecological succession).

In any case, the dilemma of 'development versus the environment' exists; sometimes it is characterized by intense antagonism among different economic and ecological priorities; it manifests itself in various places and historic periods. Characteristic examples are conflicts over the use of land between forests and pastures, wetlands and agriculture, rural and housing or touristic uses etc. The dilemma also applies to the contemporary opposition between energy abundance with nuclear technology and energy frugality with soft energy sources.

The view that the planet is threatened by ecological disasters that may lead to the extinction of mankind does not have, for the time being, significant scientific support, but the great environmental problems are undoubtedly present and threatening. Strong arguments show the following:

- Economic development has limitations set by the finite Earth environment as natural resources and as a pollution receptor.
- As long as mankind remains confined in the space of its planet, the limits of material development seem unavoidable.
- Limitations to the increase of the human population and the per capita consuming level are also unavoidable.
- The non-material dimensions of development are less influenced by environmental limitations. Telecommunication, education, computing possibilities, information and knowledge, and intellectual recreation are examples of dimensions that are not great consumers of natural resources.

An integrated approach of any kind of development's limits is also connected with investigation of the limits of man himself, of the psychological boundaries of his personality and his boundaries as a social being. Such a treatment of the issue, though, which belongs to the fields of anthropology and philosophy, does not enter common discussions about environment and development.

13.3 SUSTAINABLE DEVELOPMENT

Scientific research has offered an abundance of data that show increasing environmental degradation. These findings have contributed to the gradual realization that economic development has some limits determined by the finite environment of the planet. Such analyses, which are modern versions of Malthusian views, have created reservations versus trust of scientific and technical progress. The belief that everything is technologically possible began to be questioned when the limited endurance of the natural cycles was realized as well as the imponderable hazards from the side effects of technology. The presence of boundaries to the magnification of human

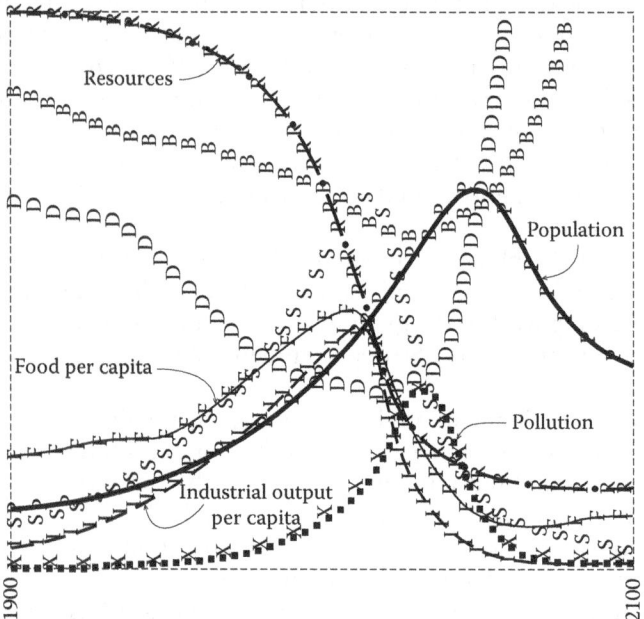

Figure 13.1 The predictions of the model on which the Club of Rome estimations were based. Since then various technological innovations and policy responses have changed some of the givens. (From Meadows, D., J. Randers, and W. Behrens (1972). *The Limits to Growth. Report of the Club of Rome.* Universe Books, New York. With kind permission from D. Meadows.)

activity was supported with some objective arguments by the Club of Rome study in 1971, in relation to the world population, natural resources and pollution indices (Figure 13.1). The UN Convention for the Environment in Stockholm (1972) was the starting point for a series of activities aiming at protection of the world environment. However, the environmental policies of various countries are not all the same, and are closely related to the level of economic development and environmental awareness.

Given the international financial and ecological mutual influence, the question of whether we should perhaps be somewhat poorer and have a better environment is meaningful at an international level. The basic developmental trend cannot be changed, since a developed economy satisfies not only the need for abundant material goods but also offers better possibilities for public health and education, recreation and security, which are all preconditions for a smooth democratic life. The great majority of the world population expects such material and intellectual goods. However, changing an extravagant lifestyle, with a restraint of excessive consumption, as for instance the thoughtless use of the private car, can play a positive role in the decrease of environmental degradation, and at the same time serve as a constructive example. Human societies, though, function mainly on the basis of the interests of individuals, groups and nations that compete between themselves for more goods and power. N. Keyfitz comments that the amount of wealth that is transferable from the United States to poor countries depends on the degree at which the average American could consume as much as the average resident of a poor country while at the same time continue to produce as an American.

The solution can only be given via a great compromise, combined with the utilization of technology. Richer nations or individuals do not accept a reduction of their rate of consumption; on the other hand, it is necessary that they maintain their high productivity. However, reduction of negative consequences is possible through changes in technology and lifestyle as well as through the organization, for instance, of urban collective transports. Self-constraint, the will for change, and solidarity with the less privileged in conjunction with the development of constructive international cooperation can contribute to better social and external relations. The necessary increase of low income can be achieved not so much by the decrease of the income of the wealthy, but rather by increase of the total wealth of humanity and parallel preservation of environmental quality and natural resources by the use of successful technologies. This approach, which is called decoupling, points out that economic development is not necessarily associated with aggravation of environmental problems.

With its study *Our Common Future* (1987), the World Commission on Environment and Development highlighted sustainability as a modern answer to the problem of the material limits of economic growth. Sustainable development aims at the rational management of natural resources such that present needs are covered, especially those of the poorer social strata and of developing countries, without undermining the fulfilment of future needs. The effort for reconciliation of opposition between economic development and the environment is inspired by the idea of sustainability that springs from the rational exploitation of the forest. Timber felling can offer continuous wood production, which corresponds to the "interest" while the "capital" remains in a good condition so that its productive capacity is perpetuated. Correspondingly, the proper policy for the satisfaction of human needs requires a simultaneous increase in productive capacity, respect towards ecological limitations, expansion of the prospects for equal distribution of goods and harmonization of demographic increase with natural system capabilities.

The feasibility of sustainable development in various cases of environmental problems is not always self-evident. Generally, however, it is useful as an analysis tool because it enables a well-documented critique of specific developmental options. It is essentially an effort of incorporating the environmental dimension into developmental policies, and it entails a significant decrease in the economic growth pace. It is a vague concept,

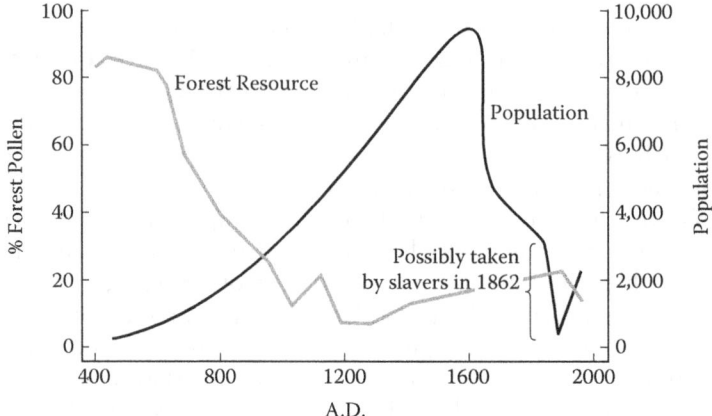

Figure 13.2 The history of the settlement of Easter Island in terms of population and forest resources. (From Bahn, P. and J.R. Flenley (1992). *Easter Island, Earth Island.* Thames and Hudson, London; and Bush, M.B. (2003). *Ecology of a Changing Planet,* 3rd edition. Prentice Hall, Upper Saddle River, NJ. With kind permission from J.R. Flenley.)

which, however, constitutes a successful slogan that has theoretically been adopted both by governments and non-governmental organizations. The emergence of interest by the international capital (green capitalism) for investments in sectors such as waste recycling, health products, renewable energy sources etc. reinforces the acceptance of relevant ideas by the international community.

Rational management often requires a cyclic movement of natural resource products either in the frame of cycles that exist in nature (renewable natural resources) or by the creation of artificial cycles (recyclable natural resources). The relative shortening of the cycle period is an important parameter for pollution avoidance (e.g. prompt recycling of used metals), while the very long period of certain natural cycles (e.g. phosphorus cycle) deprives the respective natural resource of the recyclable property. The rational management of renewable natural resources (freshwater, forest, fauna etc.) includes saving, so that exhaustion (Figure 13.2) can be avoided, as well as their protection against degradation. Rational management of non-renewable natural resources (e.g. mine products or fossil fuels) means reasoned use in conjunction with the resource's abundance level so that its medium-term exhaustion, as well as excessive pollution, can be avoided. In rational management of both renewable and non-renewable natural resource saving, reuse and recycling technologies make a significant contribution.

The exploitation manner of natural resources is directly associated with the kind of technology used. Softness, e.g. in energy technologies, is a quality that characterizes only the technologies that ensure resources' renewability and satisfactory environmental conservation and at the same time do not create social hazards (e.g. serious accidents). Of course, this last attribute allows for different interpretations, as for instance the view held by many ecological movements that the concentration of power by technocrats constitutes a social hazard. Rational management of natural resources and the use of softer technologies is to a certain extent an economic issue, because the exploitation cost is significantly increased; therefore the cost of the final product increases also, rendering development more expensive but also more environmentally friendly.

Sustainable development is characterized by the following positions and goals:

- Sustainability is inextricably associated with environmental, social and economic problems.
- Environmental pressures are interdependent, since the environment is a complex dynamic system.
- Ecological and economic problems are associated with many social and political factors, such as unemployment, poverty, social exclusion, women's inferior status in many societies etc.
- Environmental damages do not stop at national borders.
- The revitalization of the economy is necessary, especially in developing countries where poverty reduces people's ability to use natural resources prudently, and intensifies the pressures on the environment.
- A less energy consuming and more socially just type of development should be promoted.
- The increasing populations' basic needs in the developing countries, such as proper diet, necessary energy, housing, clean water, sanitary living conditions and medical care, should be satisfied.
- Employment opportunities should be created that will ensure the minimum necessary goods.
- Demographic stabilization is necessary everywhere, especially in big cities of developing countries where deficiencies in housing, water, sanitation and mass transportation are especially acute. The most important outcome of a population increase deceleration will be more livable cities.
- Natural resources, which are being depleted by the high consumption level of the industrial countries, in combination with the increasing population and consumption of the developing countries, need preservation and upgrading. The prospects of alternative solutions in agricultural production, in the technologies used and in energy are important for the decrease of atmospheric and water pollution.
- Technologies should be redefined, with a market shift towards the direction of products that are friendlier to the environment, with a longer life span, that can be recycled or that be conducive to energy saving.

- Convergence of economic and environmental goals in decision making is important; convergence can be achieved through acceptance of social responsibility and recognition of the long-term consequences of today's decisions.
- Sustainability requires wider participation in responsibilities and decisions as well as wider access of the citizens to the sources of information.

13.4 ECOLOGICAL APPROACHES OF DEVELOPMENT

The role of the social system to determination of the development model and, by extension, the environmental impacts caused, is important. Trust in the liberating prospects of scientific and technological progress has put forth the goal of limitless growth as a progressive promise. Conversely, it has been accepted by many social groups that a socialist organization of society could bring about reconciliation between man and nature. However, serious environmental problems of countries with central economic planning have been exposed and the impasse of those views has become apparent. Thus the importance of freedom, critical thinking and the essential social control of the developmental procedure has emerged. The creation of a worthwhile ecological movement in various countries is more related to their economic, cultural and political level and less to the existence of intense environmental problems.

Ecological movements have emerged in most countries; however, they appear mainly in areas of higher social development. They present a dissimilar geographic distribution and a great variety in regard to their approach to the problems, the proposals and the action forms. They constitute an imperfect combination of multiform tendencies where very different philosophical views, social notions and political ideas, from conservative or religious to extremely radical, coexist. A widely accepted fixed characteristic of those movements is local action that is expressed by the slogan "think globally, act locally"; the groups that implement it have acquired significant local roots in many developed countries.

Through the fertile activity of ecological movements as well as the multiple view-shaping processes by scientists and environmental organizations, different approaches to development and the man-nature relationship have been established:

- The most radical approaches insist on the negation of development. The rejection of antagonism, of individualism and so-called artificial needs, the emphasis on collectivity, on human communication and the satisfaction of fundamental needs would gradually lead to a society that would function according to the ecosystems' patterns. The relatively small and decentralized human societies would satisfy their needs by soft and simple technology in harmonious coexistence with nature. These radical approaches include various philosophical views that have developed in recent decades. One view is the "Deep Ecology" view, that focuses on the search for the true nature of man in relation to the natural world. It propounds the utmost eco-centrism, seeing man as part of an organic whole with the total of past, present and future nature. A holistic idea that has gained wide publicity is the "Gaia" theory, the assumption that the planet is a kind of "super-organism" whose functions are self-regulated leading to homeostasis, that is, maintenance of steady and optimal conditions for survival.

- Moderate ecological approaches are attracted by the model of eco-development. This espouses the view that conservation of renewable natural resources and environmental quality will be achieved by soft technologies, recycling, abolition of squandering and useless or harmful production, and promotion of durable products instead of single-use ones. Development per se is not rejected, but only the continuous quantitative growth that underestimates the importance of the quality of life. Ecological development aims at the most harmonious possible coexistence with nature as well as human prosperity through ecologically accepted activities.

- The most development-oriented environment-friendly approaches emphasize environmental protection and biosphere preservation. The view is that development is acceptable when it is controlled to avoid danger of causing major perturbations. The creation of proper tools, such as the principle of the polluter's responsibility, the principle of prevention and studies of the environmental impacts of various developmental activities, constitute a basic means for the necessary shift towards more environment-friendly activities. The proposals for sustainable development converge to this approach.

Differentiation between the various approaches is not always easy, and there are convergences in some issues. In relation to, e.g. the energy problem, the radical ecological view asks for man's complete integration to the natural cycles of renewable energy sources by the use of the simplest possible technology and does not pursue energy abundance; the sustainable development approach would be more interested in the avoidance of impacts and would not reject the pursuit of abundant and cheap energy if it could be produced in an environmentally acceptable way; the moderate ecological view would support intermediate stands, agreeing, e.g. with the sustainable development approach about the acceptance of natural gas use as a substitute for other more pollution-generating fuels, but condemning, along with the radical approach, any kind of nuclear energy use, including nuclear fusion.

13.5 FUTURE PROSPECTS

During and after the 1970s, there was a barrage of ominous predictions that converged to the estimation that in the beginning of the 21st century, the planet's condition would not be livable, that environmental doom was likely. These predictions have not been verified because the complexity of the systems, the abilities of technology to widen the limits, and also the social adaptation of the development process due to negative feedback from the Malthusian concerns had not been taken into consideration. Moreover, the holistic approaches that have been developed are strongly related to metaphysical views; they lie at the opposite end of an integrated systemic approach and they underestimate the systemic nature of the phenomena. The linear relation between development and environmental burdening (consumption of natural resources, pollution) is not stable, mainly because society does not stand like an immobilized spectator but plays an active role, and technology brings about significant structural changes. In the case of human population, carrying capacity changes because it is socially determined. Thus carrying capacity should not be considered as an insurmountable limit for human activities; it is significantly connected with the degree of development of productive forces, especially technology.

Civilization would not be possible without man's interventions on the environment. However, the industrial era is associated with a culmination of man's pressure upon nature; continuation of an indiscriminate developmental course maximizes the risk of multiple destabilization of the biosphere. The prospect of a radical change in economic and social human behaviour does not seem realistic. However, there is an emerging tendency on the part of more developed societies towards detachment from super-consumerism as well as towards putting more emphasis on collaboration and solidarity. The development of environmental awareness and the creation and expansion of an advanced environmental policy in developed countries are characteristic examples. Birth control measures and the demographic transition that follows financial and cultural development may lead to stabilization of the global population to tolerable levels. The prudent use of technology can offer a satisfactory consumption level along with parallel coping with environmental and natural resource degradation.

With the advent of the third millennium, the entry to a "meta-industrial era" that is already visible, if not tangible, will probably be associated with a prevalence of the reaction against environmental degradation. Environmental protection is an increasingly important scientific and technological goal; social interest in the environment is rapidly increasing and the environmental dimension is penetrating deeply into economy and the various policies. In our days and in the more developed countries, the fundamental values of freedom, equality and social solidarity that gradually led to the Rule of Law and the Welfare State tend to be enriched

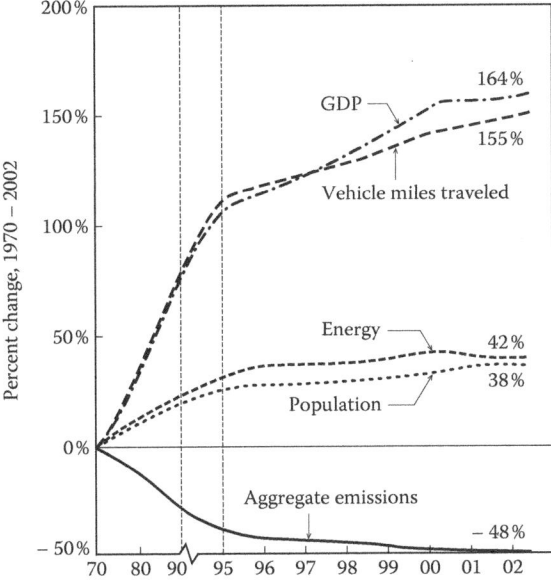

Figure 13.3 Emissions reductions of an aggregate of six principal pollutants compared with growth in GDP, vehicle miles traveled, energy, and population. (From US EPA, *Air Trends Report*, 2003, from Masters, G.M. and W.P. Ela (2008). *Introduction to Environmental Engineering and Science*, 3rd edition. Pearson, Prentice Hall, Upper Saddle River, NJ.)

with the consideration of the man-nature relationship and lead to an "Environmental Protection State". The material base of this development is already being created in an accelerating pace through new scientific and technological products that are oriented towards a friendlier to the environment production method. Today, the idea, connected with the old technological optimism, that anything is possible if enough energy and technology are available, can partly rest on existing possibilities of decoupling environmental burdening from economic growth (Figure 13.3). Modern technology presents increasing possibilities for cleaner production of goods, new agricultural production methods, pollution abatement, reuse and recycling of waste, substitution of dangerous or toxic products, energy, water, and other natural resource saving, decrease of unnecessary transportation due to computer science and telecommunications, prevention of or dealing with environmental impacts, new environmental monitoring methods, application of efficient environmental management tools etc. Spearheads towards this direction are renewable energy sources which could cover up to 100% of power production in the medium term, as well as non-material development. The future stability of the biosphere now significantly depends on the development of relevant technological and social initiatives.

Selected Bibliography

Agency for Toxic Substances and Disease Register (2012). *List of the 18 most dangerous substances in the year 2011*. www.atsdr.cdc.gov/

Assadourian, E. and M. Renner (2012). *State of the World 2012 Moving Towards Sustainable Prosperity*. The WorldWatch Institute, Washington, DC.

Bahn, P. and J.R. Flenley (1992). *Easter Island, Earth Island*. Thames and Hudson, London.

Baumann, H. (1993). *Greek Wild Flowers and Plant Lore in Ancient Greece*. Herbert Press, London.

Beilke, S. (1985). *Acid Deposition*. Commission of the European Communities, Brussels.

Bell, M. and J.C. Walker (2005). *Late Quaternary Environmental Change,* 2nd edition. Pearson, Prentice Hall, Upper Saddle River, NJ.

Berglund, B., T. Lindvall and D.H. Schwela (1999). *Guidelines for Community Noise*. World Health Organization, Geneva.

Bertalanffy, L.V. (1973). *General System Theory*. Penguin, New York.

Bird, E.C.F. (1969). *An Introduction to Systematic Geomorphology: Coasts,* vol. 4. MIT Press, Cambridge, MA.

Bragdon, C.R. (1971). *Noise Pollution*. University of Pennsylvania Press, Philadelphia.

Brown, L. (ed.) (1988). *State of the World 1988*. The WorldWatch Institute, Washington, DC.

Brown, L. (ed.) (1994). *State of the World 1994*. The WorldWatch Institute, Washington, DC.

Brown, L. (ed.) (1995). *State of the World 1995*. The WorldWatch Institute, Washington, DC.

Brown, L. (ed.) (1996). *State of the World 1996*. The WorldWatch Institute, Washington, DC.

Brown, L. (ed.) (1997). *State of the World 1997*. The WorldWatch Institute, Washington, DC.

Brown, L. (ed.) (1998). *State of the World 1998*. The WorldWatch Institute, Washington, DC.

Buclet, N. and O. Godard (eds.) (2000). *Municipal Waste Management in Europe*. Kluwer Academic, London.

Burton, J.A. (1976). *The Living Earth: Nature in the City*. Aldus Books, London.

Bush, M.B. (2003). *Ecology of a Changing Planet,* 3rd edition. Prentice Hall, Upper Saddle River, NJ.

Cancela da Fonseca, J.P. (1977). Quelques remarques à propos de la résilience et de la stabilité dans les ecosystèmes. *Revue des Questions Scientifiques*, 148: 423–444.

Carson, R. (1962). *Silent Spring*. Houghton Mifflin, Boston.

Christoulas, D., K. Hadjibiros and A. Andreadakis (1995). *Ecology Courses*, 4th edition, National Technical University of Athens, Athens.

Coase, R.H. (1960). The problem of social cost. *Journal of Law and Economics*, 3: 1–44.

Colinvaux, P. (1993). *Ecology 2*. Wiley, New York.

Cronk, Q.C.B. (1997). Islands: Stability, diversity, conservation. *Biodiversity and Conservation*, 6: 477–493.

De Angelis, D.L. (1992). *Dynamics of Nutrient Cycling and Food Webs*. Chapman & Hall, London.

Deevey, E.S. (1960). The human population. *Scientific American*, 209(9).

Department of the Environment, Transport and the Regions (2000). *Public Participation in Making Local Environmental Decisions*. London.

Detwyler, T. (1971). *Man's Impact on Environment*. McGraw-Hill, New York.

Diamond, J. (1997). *Guns, Germs, and Steel: The Fates of Human Societies*. W.W. Norton, New York.

Diekmann, A. and A. Franzen (1999). The wealth of nations and environmental concern. *Environment and Behaviour*, 31: 540–549.

Efthymiopoulos, I. and M. Modinos (eds.) (2002). *Globalization and Environment*. Ellinica Grammata, Interdisciplinary Institute for Environmental Research, Athens (in Greek).

Efstratiadis, A. and K. Hadjibiros (2011). Can an environment-friendly management policy improve the overall performance of an artificial lake? Experience from a multipurpose dam in Greece. *Environmental Science and Policy*, 14(6): 1151–1162.

Ehrlich, P.R. and A.H. Ehrlich (1970). *Population, Resources, Environment: Issues in Human Ecology*. Freeman, San Francisco.

Engelman, R., M. Renner, J. Sawin, and A. Chawla (eds.) (2009). *State of the World 2009: Into a Warming World*. The WorldWatch Institute, Washington, DC.

Fantechi, R. and A. Ghazi (eds.) (1988). *Carbon Dioxide and Other Greenhouse Gases: Climatic and Associated Impacts*. Commission of the European Communities, Kluwer Academic, Dordrecht.

Fantechi, R. and N.S. Margaris (eds.) (1986). *Desertification in Europe*. Commission of the European Communities, Reidel, Dordrecht.

Finerty, J.P. (1980). *The Population Ecology of Cycles and Small Mammals*. Yale University Press, New Haven, CT.

Flavin, C. (ed.) (2002). *State of the World 2002*. The WorldWatch Institute, Washington, DC.

Franzen, A. 2003. Environmental attitudes in international comparison: An analysis of the ISSP surveys 1993 and 2000. *Social Science Quarterly*, 84(2): 297–308.

Grossman, G.M. and A.B. Krueger (1995). Economic growth and the environment. *Quarterly Journal of Economics*, 2(110): 353–377.

Hadjibiros, K. (1981). Simulation of a predator-prey system using a matrix model. *Ecological Modeling*, 12: 45–67.

Hadjibiros, K. (1996). Sustainable development in a country with extensive presence of valuable biotopes. *The Environmentalist*, 16: 3–8.

Hadjibiros, K., P.S. Economidis, and T. Koussouris (1997). Let the fish speak: The ecological condition of major Greek rivers and lakes in relation to environmental pressures (national paper), *4th EURAQUA Technical Review*, Koblenz, pp. 103–123.

Hadjibiros, K. (2003). Landscape conservation in Greece. Inventorying sites of natural and cultural interest (national paper), *11 Paesaggio nelle Politiche Europee*, Council of Europe and Presidenza Italian a del Consiglio dell' Unione Europea, Rome, pp. 135–142.

Hadjibiros, K. (2007). *Ecology. Ecosystems and Environmental Protection,* 3rd edition. Symmetria, Athens (in Greek).

Hadjibiros, K., P. Panagiotidis, and R. Karakatsani (2007). *Dictionary of Ecological and Environmental Terms.* Staphylides, Athens (in Greek).

Hadjibiros, K., D. Dermatas, and C. Laspidou (2011). Municipal solid waste management and landfill site selection in Greece: Irrationality versus efficiency. *Global NEST Journal,* 13(2): 150–161.

Hanisch, B. and E. Kilz (1991). Waldschaden erkennen/Monitoring of forest damage/ Reconnaitre les dommages forestiers. Commission of the European Communities, Verlag/Christopher Helm/Arts Graphiques, London.

Hansen, J.A. (1996). *Management of Urban Biodegradable Wastes.* James & James Science, London.

Hardman, D.J., S. McEldowney, and S. Waite (1993). *Pollution: Ecology and Biotreatment.* Longman, New York.

Hardoy, J. (1987). Speech in Public Hearing, Sao Paulo, 28–29 October 1985. In *World Commission on Environment and Development: Our Common Future.* Oxford University Press, Oxford, UK, p 50.

Hassel, M.P. (1978). *Arthropod Predator-Prey Systems.* Princeton University Press, Princeton, NJ.

Hinkley, A. (1976). *Applied Ecology.* Macmillan, New York.

Holliday, C., S. Schmidheiny, and P. Watts (2002). *Walking the Talk. The Business Case for Sustainable Development.* Greenleaf and Berret-Koehler, San Francisco.

Holling, C.S. (ed.) (1978). *Adaptive Environmental Assessment and Management.* International Institute for Applied Systems Analysis, Wiley, New York.

Houghton, J.T., Y. Ding, D.J. Griggs, M. Noguer, P.J. van der Linden, and D. Xiaosu (eds.) (2001). *Climate Change 2001: The Scientific Basis.* Cambridge University Press, London.

Inglehart, R. (1995). Public support for the environmental protection: Objective problems and subjective values in 43 societies. *PS: Political Science and Politics,* 28: 57–72.

Inglehart, R. (1997). *Modernization and Postmodernization: Cultural, Economic and Political Change in 43 Societies.* Princeton University Press, Princeton, NJ.

IUCN, UNEP, and WWF (1991). *Caring for the Earth. A Strategy for Sustainable Living.* Gland, Switzerland.

Jain, R.K., L.V. Urban, and G.S. Stacey. (1977). *Environmental Impact Analysis.* Van Nostrand Reinhold, New York.

Jarvis, P.J. (2000). *Ecological Principles and Environmental Issues.* Prentice Hall, Upper Saddle River, NJ.

Jerlov, N.G. (1951). Optical studies of ocean water. *Reports of the Swedish Deep Sea Expedition,* 3: 1–59.

Keyfitz, N. (1972). Population Theory and Doctrine: a Historical Survey, in W. Petersen (ed.): *Readings in Population*, MacMillan, New York.

Knill, C. and A. Lenschow (2000). *New Concepts—Old Problems? The Institutional Constraints for the Effective Implementation of EU Environmental Policy*. Earthscan, London.

Kolakowski, L. (1972). *Positivist Philosophy*. Penguin Books, London.

Kormondy, E. (1976). *Concepts of Ecology*. Prentice Hall, Upper Saddle River, NJ.

Kraemer, L. (2003). Environment related processes, in Meuleman, L. et al. (eds.) *Environmental Governance in Europe*. Lemma, Utrecht, the Netherlands, pp. 69–77.

Kraft, M.E. (2007). *Environmental Policy and Politics*, 4th edition. Pearson, Longman, London.

Krebs, C.J. (2001). *Ecology*. Benjamin Cummings, San Francisco, CA.

Lambert, W. and U. Sommer (1997). *Limnoecology, the Ecology of Lakes and Streams*. Oxford University Press, Oxford.

Lawton, J.H. and R.M. May (1995). *Extinction Rates*. Oxford University Press, Oxford.

Lenschow, A. (2002). New regulatory approaches in greening EU policies. *European Law Journal*, 8(3): 19–37.

Lieth, H. and R.H. Whittaker (1975). *Primary Productivity of the Biosphere*. Springer-Verlag, Heidelberg.

Leslie, P.H. and R.M. Ranson (1940). The mortality, fertility and rate of natural increase of the vole as observed in the laboratory. *Journal of Animal Ecology*, 9: 27–52.

Levins, R. (1968). *Evolution in Changing Environments*. Princeton University Press, Princeton, NJ.

Lotka, A.J. (1925). *Elements of Physical Biology*. Williams & Wilkins, Baltimore.

Lovelock, J. (1992). *Gaia: The Practical Guide to Planetary Science*. Gaia Books, London.

Lovelock, J. (2006). *The Revenge of Gaia*. Allen Lane, London.

Malthus, T.R. (1798). *An Essay on the Principle of Population*. Johnson, London (reprinted in Everyman's Library, 1914).

Mannion, A.M. (1991). *Global Environmental Change*. Longman, New York.

Mantoglou, A., K. Hadjibiros, P. Panagopoulos, and T. Varveris (1998). Sustainable development programme for the Greek islands. *Proceedings of the International Congress on Sustainable Development in the Islands and the Role of Research and Higher Education*. Rhodes, Greece.

Margalef, R. (1983). *Limnologia, Ed.* Omega, Barcelona.

Marsh, L.L., D.R. Porter, and D.A. Salvesen (1996). *Mitigation Banking: Theory and Practice*. The Urban Land Institute, Washington, DC, Island Press.

Masters, G.M. and W.P. Ela (2008). *Introduction to Environmental Engineering and Science*, 3rd edition. Pearson, Prentice Hall, Upper Saddle River, NJ.

May, R.M. (1974). *Stability and Complexity in Model Ecosystems*. Princeton University Press, Princeton, NJ.

May, R.M. (1981). *Theoretical Ecology*. Blackwell, Oxford.

Maynard Smith, J. (1974). *Models in Ecology*. Cambridge University Press, London.

McEldowney, S., D. Hardman, and S. Waite (1993). *Pollution: Ecology and Biotreatment*. Longman, New York.

McFalls, J.A.J. (1998). Population: A lively introduction. *Population Bulletin*, 53: 1–48.

McKinney, R. (1962). *Microbiology for Sanitary Engineers*. McGraw-Hill, New York.

Meadows, D., J. Randers, and W. Behrens (1972). *The Limits to Growth. Report of the Club of Rome.* Universe Books, New York.

Mitchell, B. (2002). *Resource and Environmental Management.* Pearson, Prentice Hall, Upper Saddle River, NJ.

Modinos, M. (ed.) (2001). *Ecogeography of the Mediterranean Sea.* Stochastis/DIPE Athens (in Greek).

Mol, A.P.J. and D.A. Sonnenfeld (2000). Ecological modernization around the world: An introduction. *Environmental Politics*, 9(1): 3–16.

Mora, C. D.P. Tittensor, S. Adl, A.G.B. Simpson, and B. Worm (2011). How many species are there on earth and in the ocean? *PLOS Biology* 9(8): e1001127. doi:10.1371/journal.pbio.1001127.

Murray, R. (2002). *Zero Waste.* The Environmental Trust, Greenpeace International, Amsterdam.

Myers, N. (1979). *The Sinking Ark. A New Look at the Problem of Disappearing Species.* Pergamon Press, Oxford.

National Research Council Committee on Oil in the Sea (2002). *Oil in the sea III: Inputs, Fates and Effects.* Report. U.S. National Academy of Sciences.

Odum, E.P. (1971). *Fundamentals of Ecology.* Saunders, Philadelphia.

Odum, E.P. (1997). *Ecology. A Bridge between Science and Society.* Sinauer, Sunderland, MA.

OECD (1986). *OECD and the Environment.* Organisation for Economic Co-operation and Development, Paris.

OECD (1991). Council Recommendation on the Use of Economic Instruments in Environmental Policy, C(90) 177/final. Organisation for Economic Co-operation and Development, Paris.

OECD (2004). *Environment and Employment: An Assessment*, ENV/EPOC/WPNEP(2003)11/Final. Organisation for Economic Co-operation and Development, Paris.

O'Neil, P. (1993). *Environmental Chemistry.* Chapman & Hall, London.

O'Riordan, T. (ed.) (2000). *Environmental Science for Environmental Management.* Pearson, Prentice Hall, Upper Saddle River, NJ.

Ouvrage collectif (1977). *Encyclopédie de l'Ecologie.* Larousse, Paris.

Pielou, E.C. (1969). *An Introduction to Mathematical Ecology.* Wiley, New York.

Poole, R.W. (1974). *An Introduction to Quantitative Ecology.* McGraw-Hill, New York.

Population Reference Bureau staff (2004). Transitions in world population. *Population Bulletin*, 59(1): 3–40.

Porter, M.E. and C. van der Linde (1995). Towards a new conception of the environment—Competitiveness relationship. *Journal of Economic Perspectives*, 9(4): 97–118.

Ramade, F. (1974). *Elements d'Ecologie Appliquée.* McGraw-Hill, New York.

Ramade, F. (1981). *Ecologie des Ressources Naturelles.* Masson, Paris.

Ramade, F. (1990). *Des Systèmes Méditerrenéens.* Les Fascicules du Plan Bleu, No. 3. Economica, Paris.

Revelle, C.S., E.E. Whitlatch, and J.R. Wright (2004). *Civil and Environmental Systems Engineering.* Pearson, Prentice Hall, Upper Saddle River, NJ.

Rich, L. (1973). *Environmental Systems Engineering.* McGraw-Hill, New York.

Rifkin, J. (1998). *The Biotech Century.* Tarcher/Putman, New York.

Rossis, G. (1979). Sur la définition de la biophysique. *Revue des Questions Scientifiques*, 150(3): 339–355.

Rossis, G. (1986). *Philosophy of Life and Health Sciences*. Tsapepas, Athens (in Greek).

Ruttner, F. (1963). *Fundamentals of Limnology*. University of Toronto Press, Toronto, Canada.

Sayer, C. and P. McCarty (1978). *Chemistry for Environmental Engineering*. McGraw-Hill, New York.

Shepard, F.P. (1977). *Geological Oceanography*. Crane, Russak, New York.

Schumacher, E.F. (1977). *Small Is Beautiful*. Harper & Row, New York.

Smith, M.A., L.D. Grant, and A.I. Sors (1989). *Lead Exposure and Child Development*. Commission of the European Communities and Environmental Protection Agency, Kluwer, Dordrecht.

Solbrig, O.T., H. M.v.Emden, and P.G.W.J.v. Oordt (eds.) (1992). *Biodiversity and Global Change*. International Union of Biological Sciences, Paris.

Steinfeld, C. and D. Del Porto (2007). *Reusing the Resource: Adventures in Ecological Wastewater Recycling*. EcoWaters, New Bedford, MA.

Susskind, L.E. and W. Moomaw (eds.) (1991). *Papers on International Environmental Negotiation*, Vol. 14. PON (Program on Negotiation) Books, Harvard Law School, Cambridge, MA.

Syndicat CFDT (1980). *Le Dossier Electronucléaire*. Editions du Seuil, Paris.

Taylor, G.R. (1970). *The Doomsday Book*. Thames & Hudson, London.

Thurman, H.V. (1983). *Essentials of Oceanography*. Bell & Howell, Columbus, OH.

Treweek, J. (1999). *Ecological Impact Assessment*. Blackwell, Oxford.

Tsouni, A., N. Zervos, K. Hadjibiros, and A. Andreadakis (2002). An environmental database for the status of freshwater in Greece. *Global NEST International Journal*, 4(1): 1–14.

United Nations Environment Programme (UNEP) (1994). *Montreal Protocol on Substances that Deplete the Ozone Layer, Scientific Assessment of Ozone Depletion*. World Meteorological Organization Global Ozone Research and Monitoring Project—Report No. 37, Geneva.

United Nations Population Division (2003). *World Population Prospects: The 2002 Revision*. New York.

United Nations Population Fund (1990). *Meeting the Population Challenge*. New York (www.unfpa.org).

US EPA (2003). *Air Trends Report*. Washington, DC.

US EPA (2006). Municipal Solid Waste in the United States: 2005 Facts and Figures, Office of Solid Waste, EPA 530-R-011. Washington, DC.

U.S. National Oceanic & Atmospheric Administration (1992). Earth System Research Laboratory, http://www.esrl.noaa.gov/gmd/ccgg/trends/index.html

Usher, M.B. (1973). *Biological Management and Conservation*. Chapman & Hall, London.

Vaccari, D.A., P.F. Strom, and J.E. Alleman (2006). *Environmental Biology for Engineers and Scientists*. Wiley, New York.

Valavanides, A.P. (1995). *Chemical Factors in the Work Environment*. Synchrona Themata, Athens (in Greek).

Van Lang, A. (2002). *Droit de l' environnement*. PUF, Paris.

Warrick, R.A., E.M. Barrow, and T.M.L. Wigley (1990). *The Greenhouse Effect and Its Implications for the European Community*. Commission of the European Communities, Brussels.

Watt, K.E.F. (1973). *Principles of Environmental Science*. McGraw-Hill, New York.

Wayne, R.P., I. Barnes, P. Biggs, J.P. Burrows, C.E. Canosa-Mas, J. Hjorth, G. Le Bras, G.K. Moortgat, D. Perner, G. Poulet, G. Restelli, and H. Sidebottom (1990). *The Nitrate Radical: Physics, Chemistry and the Atmosphere*, Commission of the European Communities Air Pollution Research Report 31, Brussels.

Weidner, H. and M. Janicke (eds.) (2002). *Capacity Building in National Environmental Policy*. Springer, Heidelberg.

Wiener, N. (1948). *Cybernetics or Control and Communication in the Animal and the Machine*. Hermann and Cie, Paris, and MIT Press, Cambridge, MA.

Williamson, B. (2008). *Zero Greenhouse Emissions*. AEG, New York.

Williamson, M.H. (1972). *The Analysis of Biological Populations*. Edward Arnold, London.

Wilson, O.E. (ed.) (1988). *Biodiversity*. Smithsonian Institution, Washington, DC.

Woodwell, G.M. (1967). Toxic substances and ecological cycles. *Scientific American*, 3: 24–31.

World Commission on Environment and Development (1987). *Our Common Future*. Oxford University Press, Oxford.

Xepapadeas, A. (2005). Economic growth and the environment. In K.G. Maler and J. Vincent (eds.), *Handbook of Environmental Economics*. Elsevier, New York, pp. 1219–1271.

Zolotas, X. (1981). *Economic Growth and Declining Social Welfare*. Bank of Greece, Athens.

WEBSITES

ATSDR: Agency for Toxic Substances and Disease Registry
www.atsdr.cdc.gov

Climate Change Challenge
http://mudancasclimaticas.cptec.inpe.br/~rmclima/pdfs/destaques/sternreview_report_complete.pdf
https://www.oecd.org/env/cc/g20-climate/synthesis-investing-in-climate-investing-in-growth.pdf
www.nerc.ac.uk/about/consult/debate/climatechange/summary.asp

Environmental Legislation of the European Union
http://ec.europa.eu/environment/legal/implementation_en.htm

European Environment Agency
www.eea.europa.eu

Greenpeace
www.greenpeace.org/

Heidelberg Appeal
http://en.wikipedia.org/wiki/Heidelberg_Appeal

Landscape Impact Assessment
http://itia.ntua.gr/~kimon/Landscape_Road.pdf

NASA Earth Observatory
    http://earthobservatory.nasa.gov/
OECD
    www.oecdobserver.org/news/fullstory.php/aid/453/Decoupling_environment_
    from_economic_growth.html
Policy Development on a Coastal Landscape
    http://itia.ntua.gr/~kimon/Schinias.pdf
River Diversion
    http://itia.ntua.gr/~kimon/ACHELOOS.pdf
Sites of Outstanding Natural Beauty
    http://itia.ntua.gr/~kimon/SONB_Gr.pdf
The Right to Quiet Society
    www.quiet.org
U.S. Environmental Protection Agency
    http://epa.gov/

Appendix

The mathematical exploration of a system's stability can be carried out by the study of equations of the following type:

$$X_{t+1} = f(X_s, X_t)$$

where X_t is the value of a variable in time t, X_{t+1} is the corresponding value in time $t + 1$ (that is, after a time unit that is arbitrarily determined according to the problem), and X_s is its value at the point of equilibrium. It is permissible to assume that there is a linear relation between the successive values of the distance from the equilibrium point, when this distance is small enough, in which case the following will apply:

$$\frac{X_{t+1} - X_s}{X_t - X_s} = K \qquad X_{t+1} = X_s + K(X_t - X_s) \tag{A.1}$$

where K is stable. If $X_t = X_s$ then $X_{t+1} = X_t$, that is, X_s is indeed a point of equilibrium.

Setting $Z_t = X_t - X_s$ (Z_t measures the distance from the equilibrium point), we have:

$$Z_{t+1} = KZ_t \tag{A.2}$$

Based on (A.2) we can explore the system's behaviour types for small shifts from the equilibrium point:

If $K < -1$, then $|Z_{t+1}| > |Z_t|$ and sign $(Z_{t+1}) \neq$ sign (Z_t). The system tends to depart from the equilibrium point (instability) with deviating oscillations (Figures A.1 and A.2).

If $-1 < K < 0$, then $|Z_{t+1}| < |Z_t|$ and sign $(Z_{t+1}) \neq$ sign (Z_t). The systems tend to approach the equilibrium point (stability) with converging oscillations (Figures A.3 and A.4).

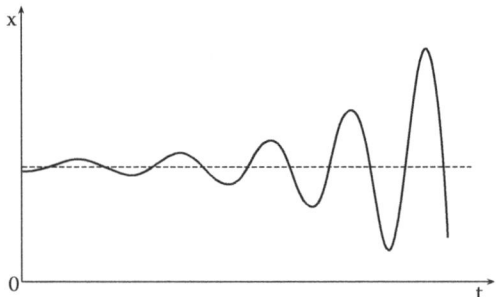

Figure A.1 Deviating oscillation. (From Hadjibiros, K. (2007). *Ecology. Ecosystems and Environmental Protection,* 3rd edition. Symmetria, Athens (in Greek). With permission.)

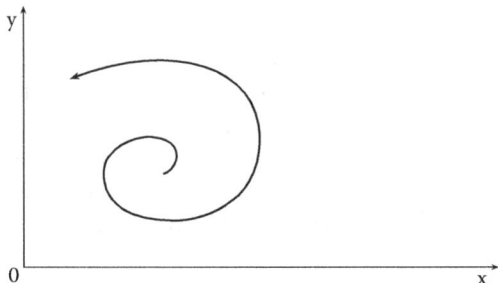

Figure A.2 Deviating oscillations in the phase space. (From Hadjibiros, K. (2007). *Ecology. Ecosystems and Environmental Protection,* 3rd edition. Symmetria, Athens (in Greek). With permission.)

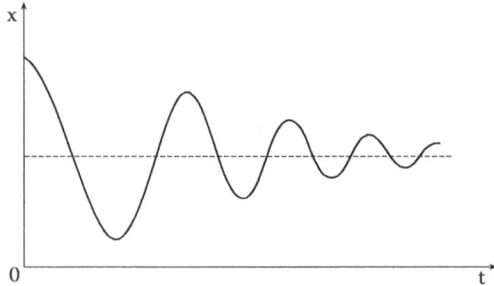

Figure A.3 Converging oscillation. (From Hadjibiros, K. (2007). *Ecology. Ecosystems and Environmental Protection,* 3rd edition. Symmetria, Athens (in Greek). With permission.)

Figure A.4 Converging oscillations in the phase space. (From Hadjibiros, K. (2007). *Ecology. Ecosystems and Environmental Protection,* 3rd edition. Symmetria, Athens (in Greek). With permission.)

If $0 < K < +1$, then $|Z_{t+1}| < |Z_t|$ and sign (Z_{t+1}) = sign (Z_t). The system tends to approach the equilibrium point (stability) with exponential convergence (Figure A.5).

If $K > +1$, then $|Z_{t+1}| > |Z_t|$ and sign (Z_{t+1}) = sign (Z_t). The system tends to depart from the equilibrium point (instability) with exponential deviation (Figure A.6).

Those conclusions result from the study of only one variable, but they are also valid for complex systems with n variables that undergo small perturbations. We should point out the value of a linear approach for small perturbations, given that according to mathematical analysis it is often acceptable as a first approach when we deal with continuous differentiable functions.

A system's response to an external perturbation can play a significant role in the regulation and control of its behaviour. The case of negative feedback is of special interest for cybernetics; in this case, the external perturbation

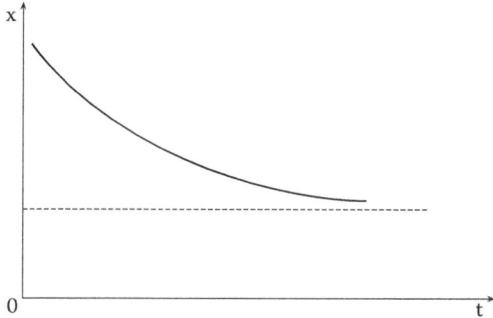

Figure A.5 Exponential convergence. (From Hadjibiros, K. (2007). *Ecology. Ecosystems and Environmental Protection,* 3rd edition. Symmetria, Athens (in Greek). With permission.)

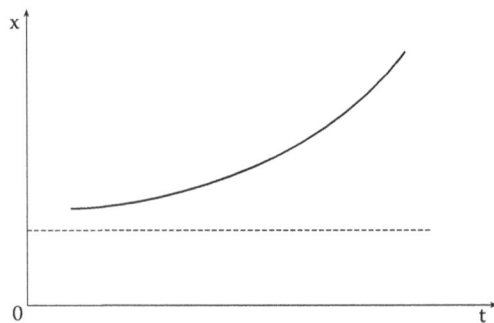

Figure A.6 Exponential deviation. (From Hadjibiros, K. (2007). *Ecology. Ecosystems and Environmental Protection,* 3rd edition. Symmetria, Athens (in Greek). With permission.)

leads the system to convergence towards the equilibrium point, which is something that serves the control. In the case of positive feedback, the external perturbation leads the system to an explosive change, that is, out of control. Negative feedback plays a fundamental role in the regulative processes of many ecological, social or other systems.

The above four behaviour types correspond to small enough perturbations so that the behaviour of the system may be considered linear. In the more general case that the system's behaviour cannot be described by linear equations, we have more complicated models, among which the *limit-circle* is of special interest (Figure A.7). In this case instead of an equilibrium point there is, in a two-dimensional phase space, a closed curved line, not necessarily circular (usually an ellipse), called the limit-circle. The system's state, when perturbed, tends to return to a point of the closed line (stability; Figure A.8) or tends to depart from it (instability). In a variables-time diagram, the limit-circle corresponds to variables' oscillations with a fixed width (Figure A.9).

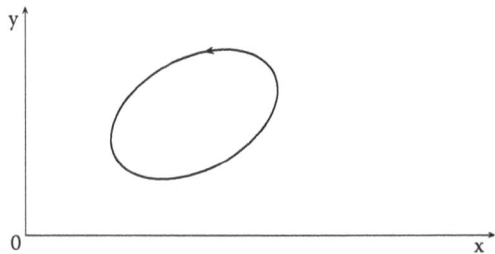

Figure A.7 Limit-circle in the phase space. (From Hadjibiros, K. (2007). *Ecology. Ecosystems and Environmental Protection,* 3rd edition. Symmetria, Athens (in Greek). With permission.)

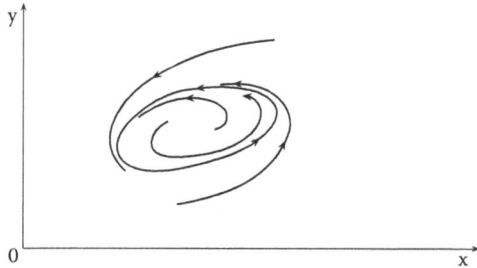

Figure A.8 Limit-circle stability. (From Hadjibiros, K. (2007). *Ecology. Ecosystems and Environmental Protection,* 3rd edition. Symmetria, Athens (in Greek). With permission.)

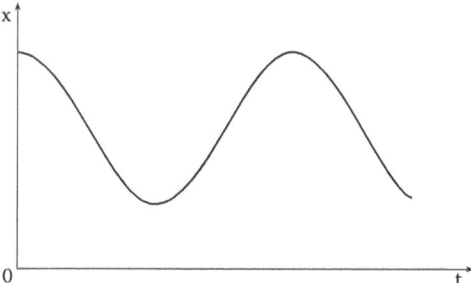

Figure A.9 Fixed width oscillation. (From Hadjibiros, K. (2007). *Ecology. Ecosystems and Environmental Protection,* 3rd edition. Symmetria, Athens (in Greek). With permission.)

If there are three variables, the limit-circle becomes limit-sphere of the three-dimensional phase space, while for n variables corresponding figures in corresponding n dimensional phase spaces will result. Thus a system can display a type of stable equilibrium even if none of the variables exhibit numerical stagnation, that is, it does not have an equilibrium point.

Cyclic behaviour may also be observed in a special case where the system does not exhibit stability or instability but is in an indifferent equilibrium. This is the so-called conservative oscillation, that is, an oscillation with a stationary width that depends on the initial conditions. Any change in the variables' value causes a permanent change to the oscillation width, or in other words, a permanently smaller or greater circle in the phase space (Figure A.10). In the extreme case that the circle becomes as small as to be a point, that is, that the variables have temporally stationary values without oscillations, the conservative oscillation ends up being a simple equilibrium point. A classic example of conservative oscillation is the mutual influence of two predator-prey populations without any self-limitation of either (Figure 3.7).

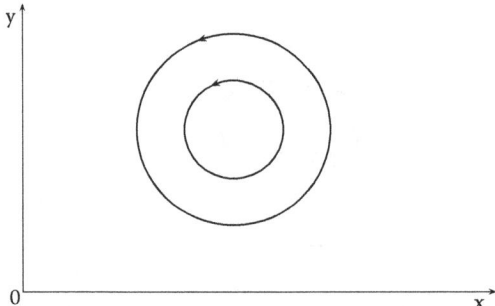

Figure A.10 Conservative oscillation. (From Hadjibiros, K. (2007). *Ecology. Ecosystems and Environmental Protection,* 3rd edition. Symmetria, Athens (in Greek). With permission.)

Index